Manual of Water Supply Practices

M19

# Emergency Planning for Water and Wastewater Utilities

Fifth Edition



American Water Works Association

Manual of Water Supply Practices — M19, Fifth Edition

# Emergency Planning for Water and Wastewater Utilities



Managing Editor: Melissa Valentine
Sr. Specialist, AWWA Manuals: Sue Bach
Cover Design: Melanie Yamamoto

---

**Library of Congress Cataloging-in-Publication Data**
Names: Gay, Stephen D., author. | Borman, Scott D., author. | American Water Works Association, issuing body.
Title: M19 emergency planning for water utilities / by Stephen D. Gay and Scott D. Borman.
Other titles: Emergency planning for water utilities | AWWA manual ; M19.
Description: Fifth edition. | Denver, CO: American Water Works Association, [2018] | Series: AWWA manual ; M19 | Includes bibliographical references.
Identifiers: LCCN 2018014152 | ISBN 9781625762795
Subjects: LCSH: Waterworks--Management. | Emergency management. | Emergency water supply.
Classification: LCC TD487 .G39 2018 | DDC 363.6/10684--dc23
LC record available at https://lccn.loc.gov/2018014152

Printed in the United States of America

ISBN 978-1-62576-279-5 eISBN 978-1-61300-460-9



American Water Works Association

American Water Works Association
6666 West Quincy Avenue
Denver, CO 80235-3098
**awwa.org**

# Contents

# Figures

This page intentionally blank.

# Tables

This page intentionally blank.

# Preface

Water and wastewater utilities, hereafter referred to as the *utility*, are responsible for providing adequate supplies of safe drinking water and for reliably collecting and treating wastewater. This manual addresses best practices for the prevention, mitigation, response, and recovery of utility operations during critical incidents and is intended for use by utilities of any size. The subject of this manual is closely tied with the AWWA management standards ANSI/AWWA G430, Security Practices for Operation Management; ANSI/AWWA G440, Emergency Preparedness Practices; and ANSI/AWWA J100, Risk and Resilience Management of Water and Wastewater Systems. This manual provides the reader with useful information and resources for satisfying the requirements set forth in the standards and related resources. Where applicable, the requirements of these standards are referenced in text box.

This fifth edition of AWWA Manual M19, *Emergency Planning for Water and Wastewater Utilities,* has been updated to align with current emergency planning best practices. Throughout the manual, the reader will find references to resources designed to support risk assessments as well as to develop mutual aid opportunities and mitigation activities.

This manual comprises the following chapters that discuss proven methodologies and best practices used by water utilities in emergency planning:

Chapter 1, Preparedness Culture, discusses how a preparedness culture increases the effectiveness of utility response to incidents as well as best practices for developing and promoting this response.

Chapter 2, Risk and Resilience Assessment, describes how water utility leaders can better understand risks to their mission and critical assets, how this assessment supports emergency planning by identifying and prioritizing utility-specific risks, and how to identify investments that can pay the greatest dividends during and after an incident.

Chapter 3, Developing an Emergency Response Plan, discusses the basic principles and elements of a preparedness plan, and highlights plan development, emergency procedures, organizational use of an incident command system, and maintenance.

Chapter 4, Mutual Aid and Partnerships, discusses the value in developing partnerships between utilities and other organizations and describes in detail the Water/Wastewater Agency Response Network (WARN), which provides mutual aid and assistance.

Chapter 5, Internal and External Communications, discusses the various modes and messages of communication with stakeholders that are critical to utility emergency response.

Chapter 6, Training and Exercises, describes how utility personnel can practice preparing for and responding to incidents to build skills that will minimize losses and expedite recovery.

Chapter 7, Mitigation, describes how to increase a utility's resilience and preparedness through risk-management activities including hazard mitigation.

This page intentionally blank.

# Acknowledgments

The AWWA M19 Emergency Planning for Water Utilities Manual Subcommittee, which developed this manual, had the following personnel at the time:

Stephen D. Gay, *Chair*

S. Bach, Staff Advisor, AWWA, Denver, Colo.
D.S. Borman, Benton/Washington Regional Public Works Authority, Rogers, Ark.
R. Bova, Suffolk County Water Authority, Oakdale, N.Y.
J. Cassidy, Critical Situation Management Inc., Marysville, Pa.
J. Crisologo, California Department of Public Health, Glendale, Calif.
S.A. Dennis, Alameda County Water District, Fremont, Calif.
D. Flancher, Standards Engineer Liaison, AWWA, Denver, Colo.
S.D. Gay, City of Westminster, Westminster, Colo.
M.F. Goddard, Seattle Public Utilities, Seattle, Wash.
D. Goldbloom-Helzner, US Environmental Protection Agency, Washington, D.C.
C. Grieves, Baxter & Woodman Inc., Crystal Lake, Ill.
M. Howe, Executive Director/Secretary Treasurer, Texas AWWA, Austin, Texas
J.L. Ignatowski, Ignatowski Environmental Engineering Consultant, Vienna, Va.
B. Kerr, Kerr & Associates Consulting, Victoria, B.C., Canada
H. Klein, Garden State Laboratories Inc., Hillside, N.J.
J. W. McLaughlin, Merrick & Company, Charlotte, N.C.
V. Morello, Broward County Water and Wastewater Services, Pompano Beach, Fla.
K. Morley, Manager Federal Relations, AWWA, Washington, D.C.
J.W. Moyer, AECOM, Morrisville, N.C.
S. Neumeister, Southeastern Technical Solutions, Port Saint Lucie, Fla.
K. Novick, Gradient Planning, LLC, Middletown, Conn.
J.W. Porco, Michael Baker Corporation, Pagosa Springs, Colo.
K. Raman, KUV Consultants, LLC, Phoenix, Ariz.
R.A. Riordan, City of San Ramon and CH2M, San Ramon, Calif.
L.P. Warren, Launch! Consulting, Charlottesville, Va.
J. Whitler, Bureau of Reclamation, Denver, Colo.
R. Wilson, CSC, Darien, Conn.

This manual was also reviewed and approved by the AWWA Standards Council and the Standards Committee on Emergency Preparedness Practices. The Standards Committee on Emergency Preparedness Practices had the following personnel at the time of approval:

D. Scott Borman, *Chair*

*General Interest Members*

T.P. Allman, Naval Facilities Engineering Command, Misawa, Japan
D.M. Apanian, USEPA Region 4, Atlanta, Ga.
J.E. Crisologo, State Water Resources Control Board, Glendale, Calif.
D.M. Flancher,* Standards Engineer Liaison, AWWA, Denver, Colo.
R. Ford,* Standards Council Liaison, CH2M, Parsippany, N.J.
K. Morley,* Manager Federal Relations, AWWA, Washington, D.C.
S.M. Neumeister, Southeastern Technical Solutions, Port Saint Lucie, Fla.
K. Novick, Gradient Planning, LLC, Middletown, Conn.
C. Sapp, CH2M, Sugar Hill, Ga.
G.G. Sturdivan, Sierra West Consultants Inc., Yucca Valley, Calif.
L.P. Warren, Launch! Consulting, Charlottesville, Va.

*Producer Members*

J.K. Cassidy, Critical Situation Management Inc., Marysville, Pa.
C. Herndon, Herndon Solutions Group, Las Vegas, Nev.

*User Members*

D.S. Borman, Benton/Washington Regional Public Works Authority, Rogers, Ark.
S.D. Gay, City of Westminster, Westminster, Colo.
M.F. Goddard, New Orleans Sewerage and Water Board, New Orleans, La.
H. Klein, Garden State Laboratories Inc., Hillside, N.J.
M.R. Nandagopal, City of Spokane, Spokane, Wash.
C. Perry, City of Hurst, Hurst, Texas
K. Schweitzer, Hicksville Water District, Hicksville, N.Y.
S. Stephens, Austin Water Utility, Austin, Texas
P. Thompson, Suffolk County Water Authority, Oakdale, N.Y.

* Nonvoting

Chapter 1

# Preparedness Culture

One mark of a utility that has the capacity to respond effectively to emergencies is a strong preparedness culture. This culture is evident when an emergency response plan (ERP) is kept current—staff are trained in their responsibilities as they relate to the plan; resources necessary to implement the ERP are supported by the budget; staff are aware of emergency preparedness activities and participate in them; an explicit organizational commitment to emergency planning is made; and a management process is in place for staff to take corrective action when necessary improvements are identified. Improvements are identified through risk assessments, plan updates, training, exercises, incident investigations, and other activities described in this manual. When a utility has a preparedness culture, the ERP "lives" as a resource that is adapted and updated whenever there are significant changes that affect the plan. Also, staff are empowered to speak up and constructively affect change that ultimately improves the utility's level of preparedness. With this culture in place, preparedness is not a burden but an aspect of the utility that is valued and protected by staff, leadership, and other stakeholders due to its significant benefits.

Most utilities are very good at meeting their mission during normal operations, which include small and even routine process upsets. However, when upsets become nonroutine and large in scale or when emergencies arise, utilities with strong and effective preparedness cultures are better equipped to restore the critical lifeline services they provide each community. In contrast, those utilities that lack a strong preparedness culture struggle to restore services, which in some cases exacerbates the situation.

This chapter outlines and defines the concept of "culture" within an organization and presents the core components of a strong emergency preparedness culture within a utility. Organizational culture refers to a system of shared meaning that distinguishes the organization from other organizations. This shared meaning is held by organization members. A strong culture is characterized by the organization's core values being both intensely held and widely shared (Wiener 1998). A preparedness culture, therefore, is one where being prepared for emergencies is highly valued. The core components of the preparedness culture cycle, as illustrated in Figure 1-1, include the following:

Figure 1-1 The Preparedness Culture cycle

- an explicit organizational commitment toward preparedness,
- development of the preparedness culture,
- support for the preparedness culture with tools and tactics, and
- evaluation of the "strength" of the preparedness culture.

The following sections provide examples of how a utility can address each component (Dwivedi).

> *ANSI/AWWA G440, Sec. 4.1.1: Explicit and visible commitment of senior leadership to preparedness. The utility shall establish an explicit, visible, and easily communicated commitment to emergency preparedness.*

## DEVELOPING AN EXPLICIT ORGANIZATIONAL COMMITMENT TO EMERGENCY PREPAREDNESS

The utility's general manager, public works director, or chief executive officer should demonstrate a genuine commitment to emergency preparedness. This commitment is most easily shown by participation and support for the planning effort. When senior leadership works with staff to develop and refine preparedness programs, policies, and procedures, it sends a strong message to staff, governing boards, public safety officials, and the community being served by the utility. Leading by example is a critical component of building team members' trust, support, and commitment.

The establishment of an effective preparedness culture involves the shaping of enterprise-wide core values and promoting a "norm" of universal commitment to emergency preparedness. The development of a culture of preparedness relies on a demonstrated commitment throughout the entire organization. Emergency preparedness should be elevated and acknowledged within the utility's core mission statement. Utility mission statements often include statements regarding the provision of "safe and reliable" drinking water and wastewater treatment. By definition, "safe and reliable" should include a component of emergency preparedness.

## DEVELOPING THE PREPAREDNESS CULTURE

While the path to establishing a culture of preparedness begins with a commitment from senior leadership, it also relies heavily on the actions of key program stakeholders. Those tasked with preparing and maintaining the organization's ERP play a critical role in building and nurturing a culture of preparedness. Additionally, successful achievement of a true preparedness culture requires more than leaderships' demonstration and statement of commitment. It also requires staff recognition of a genuine concern and consideration for responder safety and the safety and well-being of responders' families.

Examples of actions aimed at demonstrating and reinforcing a clear commitment to the emergency planning effort and development of a culture of preparedness include the following:

> ***ANSI/AWWA G440, Sec. 4.2.1:*** *Promote emergency preparedness awareness. The utility shall promote a culture whereby management and personnel/employees understand and contribute to enhanced utility preparedness.*

- inclusion of a statement that conveys a commitment to emergency preparedness in the organization's mission statement;
- direct participation and a verbal statement of support and agency commitment at preparedness planning meetings;
- attendance and direct participation with all staff during emergency preparedness response plan training and response exercises;
- promotion of printed materials and posted commitments of emergency preparedness throughout the organization;
- institution of a recognition award program for staff who demonstrate actions aimed at personal and family preparedness;
- promotion and attendance at annual all-staff events/workshops centered around promoting emergency preparedness and recognizing those staff who have shown an outstanding effort to embrace the culture of preparedness;
- adoption of awareness programs that are directed at employees and the general public; and
- adoption of some level of funding in the operating budget that is dedicated to supporting preparedness activities such as training and/or awards programs.

One of the most important elements in supporting and maintaining a strong preparedness culture is communication. People need to actively be reminded of the risks, purpose, and objectives of being prepared and their role in the overall plan. These messages can be delivered in many ways, including the following:

- regular emergency drills and exercises (e.g., tabletop exercises);
- annual employee preparedness fairs held onsite at the utility where the efforts to be prepared are recognized by top management and where employees can purchase discounted personal emergency preparedness supplies for themselves and their families; and

- recognition among staff for participation in emergency response activities or for actions that demonstrate they are better prepared for disaster. These include giving away t-shirts or coffee mugs that indicate the event response or preparedness actions taken.

Additional information on preparedness training and exercises can be found in Chapter 6.

Other efforts to communicate the need to prepare can be passive and done on an ongoing basis. The messaging of preparedness should be "branded" around the utility and within regular communications to staff. This can be accomplished in the following ways:

> ***ANSI/AWWA G440, Sec. 4.3.1:*** *Identify employees responsible for emergency preparedness. The utility shall identify employees responsible for establishing and maintaining a level of emergency preparedness correlated with known or perceived risks. These employees shall also become an integral part of the overall security program.*

- Posting the utility's statement of commitment to emergency preparedness (included in the mission statement) in and around both public access and employee-only areas at the utility. The more people are exposed to the official message of preparedness, the more it becomes a "norm" within the culture.
- Discussing emergency preparedness during staff meetings.
- Maintaining updated and current emergency-preparedness-themed bulletin boards around the utility.
- Single-line messages of preparedness in common areas of the utility (e.g., lunchroom, meeting rooms).
- Including a standard statement that relates the need to be prepared at the bottom or within the signature line of all e-mail messages from management to staff.

## SUPPORTING THE PREPAREDNESS CULTURE WITH TOOLS AND TACTICS

> ***ANSI/AWWA G440, Sec. 4.2.2:*** *Sustain focus on preparedness. The utility shall sustain focus on preparedness by maintaining it as a high priority.*

> ***ANSI/AWWA G440, Sec. 4.3.2:*** *Establish emergency preparedness expectations. The utility shall identify and disseminate emergency preparedness expectations for staff and periodically review employee performance and proficiencies.*

When updating an ERP, consider how the plan might support the preparedness culture. It is very important to focus the plan directly on the staff who need to be prepared. This can be done using the results of the risk assessment (see Chapter 2) to guide ERP updates on all hazards and the most vulnerable critical assets and by including policies and procedures that will make an impact in incident outcomes. Some questions to ask when updating an ERP include the following:

- Does this update support the utility's commitment to emergency preparedness?
- Does this update provide the staff with information that they will need during an emergency?
- Is this update supportive of actual conditions that exist during an emergency (e.g., recognition of the need to secure the safety of family before responding)?

The planning process should be open and transparent. Regular solicitation of comments on plan updates should be supported with scheduled meetings. It should be communicated that staff review and participation are necessary for an effective plan and are therefore expected. This supports the preparedness culture with a sense of plan ownership that extends throughout the utility. This culture may be further enhanced with the following activities:

- Leverage staff meetings to inform employees of the importance of emergency preparedness at the utility.
- Keep staff updated on milestones in emergency preparedness. One example is an emergency preparedness bulletin board, and another is inclusion of emergency preparedness as an agenda item for routine meetings.
- Manage preparedness activities through a committee or stakeholder review team composed of representatives from each aspect or department of the utility, including staff members who are most critical to emergency preparedness and response activities.
- Use small teams to manage specific preparedness activities to accomplish a specific goal such as planning for a functional exercise or updating a risk assessment.

## EVALUATING THE "STRENGTH" OF THE PREPAREDNESS CULTURE

Ongoing evaluation to measure progress and set new goals will maintain a strong preparedness culture. Remember that even the very best written ERP will fail unless it remains current and is understood in purpose and process by all who are expected to participate in preparedness and response activities. Keeping the plan current requires keeping the preparedness culture strong. Ongoing evaluations typically occur whenever a risk assessment or ERP is updated, training and exercise results are reviewed, and responses to real incidents are investigated. Each evaluation in a utility's preparedness program should be documented and include the evaluation results and recommendations for improvement. Within an active culture of preparedness, recommendations for improvement are inevitable due to the shared goal of an emergency preparedness program.

At least once per year, an objective person or group should evaluate the emergency preparedness program. The questions posed should be direct and aimed at the emergency planning effort's core elements that have a direct impact on the culture. Standard questions may include the following:

- With the emergency preparedness efforts undertaken, do you feel you understand the organization's need to be able to respond effectively?
- Do you feel as though you have been given the correct amount of information to respond effectively during an emergency?
- How can the organization better support your need to personally prepare for emergencies (e.g., family preparedness)?
- What would you like to see done to improve the utility's overall preparedness?

The information gained from regular evaluations of the emergency planning effort and the related strength of the culture should provide a guide for adopting new ideas. It is important to acknowledge those who bring forward new ideas to strengthen the plan, as is using every opportunity to recognize and boost stakeholder participation. The preparedness culture of the organization will reap the benefit.

## REFERENCES AND SUPPLEMENTAL READINGS

ANSI/AWWA G440-17. Emergency Preparedness Practices. Denver, CO: AWWA.

Wiener, Y. 1998. Forms of Value Systems: A Focus on Organizational Effectiveness and Cultural Change and Maintenance. *Academy of Management Review*, 13:4:534–535.

This page intentionally blank.

# Chapter 2

# Risk and Resilience Assessment

Utilities are critical to the protection of public health and the environment, ensuring the economic vitality and public confidence in the communities they serve. It is therefore a utility's goal to be prepared for natural, intentional, and technological threats that could adversely impact operations and service deliveries.

This chapter describes how utility leaders can better understand risks to their mission and critical assets by performing a risk and resilience assessment, how this assessment supports emergency planning by identifying and prioritizing utility-specific risks, and how to identify investments that can pay the greatest dividends during and after an incident.

## WHAT IS A RISK AND RESILIENCE ASSESSMENT?

A risk and resilience assessment is a process designed to identify hazards and their relative probabilities of occurrence, to identify assets at risk, and to quantify the potential impact of the hazards on those assets at risk. A risk and resilience assessment answers questions such as the following:

- What are the worst reasonable threats that could impact utility operations?
- What makes the utility vulnerable to various threats?
- What is the best way to allocate limited resources and strengthen the utility against the impact of an identified risk?
- How might a supply chain disruption impact utility operation?
- What interdependencies exist that, if lost, could disrupt service delivery (e.g., electrical power)?

- What new risks might a utility face with system expansion or development near utility assets?
- How will geographically relevant natural hazards affect utility operations?

## GOALS OF A RISK AND RESILIENCE ASSESSMENT

The primary goals of a risk and resilience assessment are to become more aware of the risks to which a utility is exposed and to identify options that can mitigate undesirable consequences. Utilities provide essential lifeline services to their communities. Therefore, it is imperative that utility investments are made to ensure day-to-day continuity of service during times of crisis. Identifying risk and potential mitigation strategies before an event occurs is a vital step that results in huge dividends during times of crisis.

The goal of a risk and resilience assessment is to better enable a utility to do the following:

- understand threats, consequences, and vulnerabilities that may negatively impact operations;
- inform efficient resource allocations to maximize risk mitigation and improve resiliency; and
- inform and update the emergency response plan (ERP) to enhance its effectiveness during an incident.

## MANAGING A RISK AND RESILIENCE ASSESSMENT

In coordination with experts from within the utility as well as those outside the utility (e.g., local emergency management professionals, law enforcement, consultants), a utility should form a committee to conduct and/or guide the risk and resilience assessment. This committee should include those with expertise on the system(s) being assessed, the history of emergencies and near misses that affected the utility, and each part of the utility organization. One person, assigned to lead this committee, should be responsible for ensuring adequate representation from utility staff and, as needed, outside participation in the risk and resilience assessment. Also, this person should ensure that the risk and resilience assessment is documented and make recommendations to fund key risk mitigation measures that are identified as a result of conducting the assessment.

## CONDUCTING A RISK AND RESILIENCE ASSESSMENT

The Bioterrorism Act of 2002 (42 USC 300i-2) required drinking water utilities to conduct a vulnerability assessment (now commonly referred to as a *risk assessment*). While a similar federal mandate was not placed on wastewater systems, many have performed risk assessments since it is recognized as a best practice. It is recommended that risk assessments be reviewed and updated no less than every five years per ANSI/AWWA G430, Security Practices for Operation and Management. It is also recommended that the assessment be based on the methodology in ANSI/AWWA J100, Risk and Resilience Management of Water and Wastewater Systems. ANSI/AWWA J100, which is suitable and scalable for use by utilities of any size or type, follows a seven-step process, as shown in Figure 2-1.

A risk and resilience assessment should cover all critical assets including people, equipment, systems, materials and supplies, records, and other resources critical to the utility's mission. The assessment should prioritize threats, assets, and vulnerabilities as a function of consequences and what could be lost, including life, property, consumer

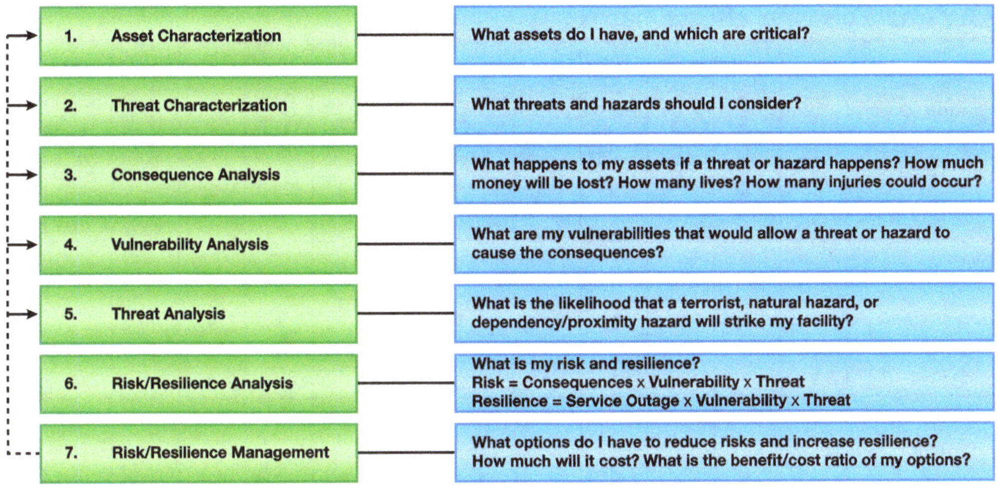


Figure 2-1 Risk assessment process as described in ANSI/AWWA J100

confidence, and continuity of service to priority users and others. With this information, utility managers can develop an action plan that can be followed to mitigate risks and update their ERP. For more information on risk and resilience mitigation strategies, see Chapter 7.

An assessment should also include mission-essential functions to support the development of a business continuity plan (see Chapter 3 for more information). Each time an assessment is revised and updated, useful information is gathered about the risks to the utility, how to mitigate those risks, and how to improve preparedness.

## THREAT CONSIDERATIONS

When conducting a risk assessment, the first step is to identify relevant threats and hazards. Information on natural, human-caused, and technological hazards is presented in subsequent sections of this chapter. It is helpful to review this information before conducting a risk assessment.

The impact of various threats and hazards on utility operation are listed in Table 2-1 and should be used to inform the assessment. ANSI/AWWA J100 includes a set of reference threats that a utility is required to consider as part of the assessment. Only threats that are relevant to the utility's location are retained for further analysis based on a worst reasonable case approach. For example, hurricanes are very unlikely to impact North Dakota, so that threat, while considered, is set aside from further analysis. Alternatively, earthquakes are a viable threat on the West Coast and are retained for further analysis. Based on the moment magnitude scale, a utility might only consider the impact of earthquakes with a maximum selected/design magnitude based on the assumption that anything larger is beyond any reasonably affordable design.

ANSI/AWWA J100 reference threats include natural hazards (e.g., earthquakes, hurricanes, tornadoes, wildfires, ice storms), dependency and proximity threats (e.g., to the power supply, transportation, supply chain), offsite threats (e.g., at industrial sites, source water or underground tanks, at a gas station), and malevolent threats (e.g., intentional contamination, cybersecurity threat, direct physical attack).

**Table 2-1 Example disaster effects matrix**

| System components—likely damage, loss, or shortage due to hazards | Earthquakes | Hurricanes | Tornadoes | Floods | Forest or Brush Fires | Volcanic Eruptions | Other Severe Weather | Waterborne Disease | Hazardous Material | Structure Fire | Construction Accidents | Transportation Accidents | Nuclear Event | Vandal, Riots, Strikes |
|---|---|---|---|---|---|---|---|---|---|---|---|---|---|---|
| Administration/operations | | | | | | | | | | | | | | |
| Personnel | ◆ | ◆ | ◆ | ◆ | ◆ | | | | ◆ | ◆ | | | ◆ | ◆ |
| Facilities/equipment | ◆ | ◆ | ◆ | ◆ | ◆ | ◆ | ◆ | ◆ | ◆ | ◆ | ◆ | ◆ | ◆ | ◆ |
| Records | ◆ | ◆ | ◆ | ◆ | ◆ | ◆ | | | | ◆ | | | | ◆ |
| Source water | | | | | | | | | | | | | | |
| Watersheds/surface sources | | ◆ | | ◆ | | ◆ | | ◆ | ◆ | | | | | ◆ |
| Reservoirs and dams | ◆ | ◆ | | ◆ | ◆ | | | ◆ | ◆ | | | | ◆ | ◆ |
| Groundwater sources | | | | ◆ | | | | ◆ | ◆ | | | | ◆ | |
| Wells and galleries | ◆ | ◆ | ◆ | ◆ | | ◆ | ◆ | ◆ | ◆ | | | | ◆ | ◆ |
| Transmission | | | | | | | | | | | | | | |
| Intake structures | ◆ | ◆ | ◆ | ◆ | | ◆ | ◆ | | | | | ◆ | | ◆ |
| Aqueducts | ◆ | | | | | | | | | | ◆ | ◆ | ◆ | |
| Pump stations | ◆ | | ◆ | ◆ | ◆ | ◆ | ◆ | | ◆ | ◆ | ◆ | ◆ | | ◆ |
| Pipelines, valves | ◆ | ◆ | | | | | ◆ | | | | ◆ | | | ◆ |
| Treatment | | | | | | | | | | | | | | |
| Facility structures | ◆ | ◆ | ◆ | ◆ | ◆ | | | | | ◆ | ◆ | | | ◆ |
| Controls | ◆ | ◆ | ◆ | ◆ | ◆ | ◆ | ◆ | | | ◆ | ◆ | ◆ | ◆ | ◆ |
| Equipment | ◆ | ◆ | ◆ | ◆ | ◆ | ◆ | ◆ | | | ◆ | ◆ | ◆ | | ◆ |
| Chemicals | ◆ | ◆ | ◆ | ◆ | ◆ | | | | | ◆ | | ◆ | ◆ | ◆ |
| Storage | | | | | | | | | | | | | | |
| Tanks | ◆ | | | | | | ◆ | | | | | ◆ | | ◆ |
| Valves | ◆ | ◆ | ◆ | | ◆ | ◆ | ◆ | ◆ | ◆ | ◆ | | ◆ | ◆ | ◆ |
| Piping | ◆ | | | | | | ◆ | | | | | | | |
| Distribution | | | | | | | | | | | | | | |
| Pipelines, valves | ◆ | ◆ | | | | | ◆ | ◆ | ◆ | | ◆ | | | ◆ |
| Pump or pressure-regulating valve stations | ◆ | ◆ | ◆ | ◆ | ◆ | ◆ | ◆ | | | ◆ | ◆ | ◆ | | ◆ |
| Materials | ◆ | ◆ | | ◆ | ◆ | | | | | ◆ | | | | ◆ |
| Electric power | | | | | | | | | | | | | | |
| Substations | ◆ | ◆ | ◆ | ◆ | ◆ | | ◆ | | | ◆ | ◆ | ◆ | ◆ | ◆ |
| Transmission lines | ◆ | ◆ | ◆ | ◆ | ◆ | | ◆ | | | ◆ | ◆ | ◆ | | ◆ |
| Transformers | ◆ | ◆ | ◆ | ◆ | ◆ | ◆ | ◆ | | | ◆ | ◆ | ◆ | | ◆ |
| Standby generators | | | | ◆ | | | | | | ◆ | | | | ◆ |
| Transportation | | | | | | | | | | | | | | |
| Vehicles | | ◆ | | ◆ | ◆ | | ◆ | | | | ◆ | ◆ | | |
| Maintenance facilities | ◆ | ◆ | ◆ | ◆ | ◆ | | ◆ | | | | | ◆ | | ◆ |
| Supplies | | ◆ | ◆ | ◆ | ◆ | | | | | ◆ | | | | ◆ |
| Roadway infrastructure | ◆ | | | ◆ | | ◆ | ◆ | | | ◆ | ◆ | ◆ | | ◆ |
| Communications | | | | | | | | | | | | | | |
| Telephone | ◆ | | ◆ | ◆ | ◆ | | | | | | | | | ◆ |
| Two-way radio | | ◆ | | | | | ◆ | | | ◆ | ◆ | ◆ | | ◆ |
| Telemetry | ◆ | | ◆ | ◆ | ◆ | ◆ | | | | | | | | ◆ |

## Natural Hazards

Natural disasters during the first decades of the 21st century were among the most severe in American history in terms of both deaths and damage. In fact, in 2011, more than $12 billion in damage and the deaths of more than 1,000 people were attributed to weather-related disasters. Many of these disasters had profound disruptive effects on utilities. In 2005, Hurricane Katrina struck the Gulf Coast, causing major aboveground damage to systems in many cities, especially New Orleans. In May 2011, more than 180 tornadoes struck the Midwest, including a devastating F5 tornado that destroyed Joplin, Missouri. The record drought in Texas and the lower Midwest significantly stressed water supplies. Although it did not happen in the United States, the 2011 Tohoku earthquake in Japan and resulting tsunami remind us of the fragility of infrastructure and the critical interdependencies among utilities.

**Earthquakes.** Earthquake damage can be widespread and severe. The energy released by an earthquake can be equivalent to the explosion of millions of tons of dynamite. The degree of resultant damage is related to the earthquake's magnitude, the distance from the epicenter (i.e., the part of the earth's surface directly above the subsurface focus of an earthquake), the soil types in the area, the resulting mode of earth failure, and the design and materials of structures.

From a practical standpoint, earthquake risk for areas of North America has been defined with relatively high precision using geological and historical information. Figure 2-2 illustrates seismic damage risk in the United States. While it is apparent that portions of the Pacific Coast have the highest risk, other areas are at significant risk. To determine a more precise probability of an earthquake occurring and the potential magnitude, utilities should review local geological and seismological data available from such sources as the US Geological Survey, state and provincial geological agencies, and universities, or the utility can employ a consultant experienced in earthquake hazard analysis.

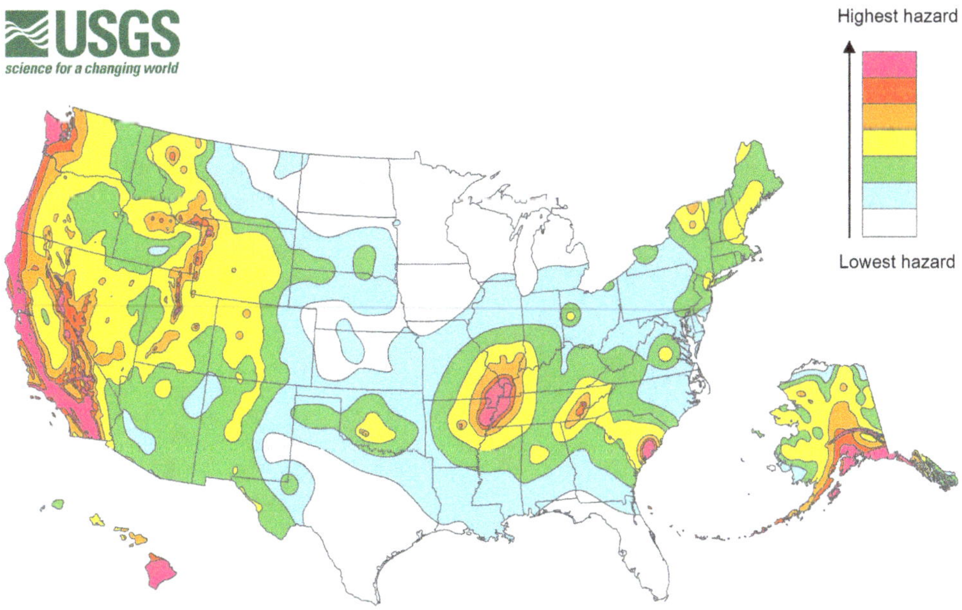


*Source:* US Geological Survey Seismic Hazard Mapping Project.

**Figure 2-2 Seismic hazard map of the United States**

*Fault rupture.* A fault rupture is the differential movement of two land masses along a fault. The 1992 Landers–Big Bear (California) earthquake created a 21-ft (7-m) offset in some areas (Figures 2-3A and 2-3B). If a pipeline lies across a fault that ruptures, it will be exposed to forces that can deform or split the pipe or even separate pipe sections at the joints.

Known faults are identified on geological and seismic maps and are often readily visible from light detection and ranging mapping, aerial photographs, and occasionally

*Source:* Photo by Michael O'Rourke. Person in picture, LeVal Lund.

**Figure 2-3A Landers–Big Bear earthquake fault offset**

*Source:* Photo by Michael O'Rourke.

**Figure 2-3B Landers–Big Bear earthquake pipe broken in the fault offset**

at ground level. If a fault crossing is required, the pipeline design should allow for fault movement. In northern California, the San Francisco Public Utilities District has implemented sophisticated pipeline designs for crossing the Calaveras and Hayward faults.

*Ground shaking.* The energy created by an earthquake radiates in waves from the epicenter like ripples that appear after a stone is thrown into a pond. The propagating waves cause the ground to shake. The closer to the epicenter, the more violent and vertical the ground shaking; conversely, the farther from the epicenter, the less severe and more rolling the ground shaking. Soil types also affect the intensity of ground shaking. Released energy produces both vertical and horizontal accelerations that can damage utility system components. For example, a pipe without flexible joints that enters a concrete structure may be subject to shear failure due to the differential dynamic response between the buried pipe and the structure. Another result of seismic forces is the rotational moment produced at the base of water storage tanks that results from internal hydrodynamic forces, as illustrated in Figure 2-4.

*Liquefaction.* The energy of an earthquake generates high pore-water pressure that can overcome the friction between soil particles. When the pressure of the water located in the spaces between soil particles, or pore water, exceeds particle friction forces, particularly in loose sands with high water content, the soil will become a soil–water slurry with significantly reduced shear strength. The result can be foundation bearing failure, differential settlement, and lateral spreading or floating of underground components. Figure 2-5A shows a ground crack caused by lateral spreading and Figure 2-5B shows the resulting pipe joint separation. Liquefaction can also cause a loss of slope stability and produce landslides.

*Densification and consolidation.* Earthquake-induced ground shaking can cause soil to become denser. The process is similar to using a vibratory roller to compact fill or to shake loose sand in a tray to make it more compact. Ground subsidence can occur as the soil densifies. In Valdivia, Chile, subsidence due to densification during a 1960 earthquake amounted to more than 3 ft (1 m) (Lambe and Whitman 1969).

*Landslides.* Landslides occur when slopes lose shear strength because of a disturbance such as ground shaking. Most damage caused by the 1964 Alaskan earthquake resulted from landslides. Landslides can cause earthen dams to fail and can also impact source

**Figure 2-4 Seismic forces on water storage tank**

*Source:* Photo by Don Ballantyne.

**Figure 2-5A Ground crack resulting from lateral spreading in Kobe, Japan in 1995**

*Source:* Photo by Don Ballantyne.

**Figure 2-5B Ductile-iron joint pulled apart due to lateral spreading**

water if large amounts of debris enter a reservoir. Landslide hazard maps have been developed for some earthquake-prone areas.

*Tsunamis and seiches.* A tsunami is a series of ocean waves generated by sudden displacements in the sea floor, landslides, or volcanic activity. In the deep ocean, a tsunami wave may be only a few inches high. A tsunami wave may come ashore gently or may increase in height to become a fast-moving wall of turbulent water several meters high. Earthquakes, volcanic eruptions, and other underwater explosions, landslides, sudden glacier activity, meteorite impact, and other disturbances above or below water have the potential to generate a tsunami. Tsunamis cause damage in two ways: the smashing force of a wall of water travelling at high speed and the destructive power of a large volume of water draining off the land and carrying everything with it, even if the wave did not look large. The wall of water from a tsunami can greatly exceed normal high-tide levels.

A seiche is an oscillation of the surface of a lake, reservoir, or landlocked sea due to an earthquake. A seiche can overtop dams, erode earthen dams, and result in dam failure. Seiches can also damage dam inlet and outlet structures.

**Wind hazards.** High winds are known to cause damage on a massive scale and often contribute to other hazards such as flooding, erosion, wildfires, structural damage, and contamination. There are many types of wind hazards including tornadoes, hurricanes and other cyclones, and straight-line wind events. These conditions can also impact the utility crew's ability to operate safely in the field.

*Tornadoes.* Tornadoes most commonly occur in the central United States; however, they have been reported in all 50 states. The most common prevalent tornado hazard is extremely high winds. Exact speeds have not been measured because wind gauges rarely survive a tornado, but estimates are as high as 300 mph (480 km/h). While tornadoes can be tracked by radar and warnings can be issued, the actual spawning of a tornado can occur with little or no warning. Because of this, tornadoes can be deadly. Structural damage from tornadoes can be extensive as aboveground facilities are rarely designed to withstand tornado-force winds.

*Hurricanes.* The costliest natural disasters in the United States in recent years have been hurricanes. Approximately 44 million people in the United States live in areas that have been hit by hurricanes. The main hazards associated with hurricanes are high winds, storm surge along shorelines, and flooding. Surface structures can suffer wind damage, and flooding can destroy plant facilities and contaminate water supplies. Structural damage and the roots of trees toppled in a hurricane can break pipelines and result in tremendous flow demands due to numerous leaks. Power outages are also widespread. Damages in the tens of billions of dollars is not uncommon.

The 2017 hurricane season ranks among the most intense and the most expensive in history. Most of the damage was caused by 3 of 17 named storms, namely, Harvey, Irma, and Maria. From Harvey's historic flooding of Texas to Irma's pounding of the Florida Keys and Maria's devastation in Puerto Rico, early estimates indicate the United States suffered more than $200 billion with 207 fatalities from the deadly trio.

The damage caused by Harvey was not as much from the Category 4 winds, but rather from a slow-moving storm that dumped an estimated 51 in. or 27 trillion gallons of rain over Texas and Louisiana in a six-day period. Florida Department of Environmental Protection reported 85 percent of drinking water and wastewater facilities were impacted by Irma, causing an estimated 500 boil-water notices and approximately 600 sewer system overflows. Federal Emergency Management Agency (FEMA) reports that progress is slow, but four months after Maria made landfall, water services are getting back on track.

In contrast, Hurricane Katrina in 2005 inflicted nearly $125 billion (2005 USD) in damage with over 1,800 deaths. Hurricane Katrina remains the single most catastrophic natural disaster in US history and is tied with Hurricane Harvey as the costliest hurricane in US history. Damage in New Orleans was exacerbated by flooding from failed levees, which inundated 80 percent of the city. There was major damage to numerous Louisiana utilities, where 23 percent of the drinking water facilities were inoperable soon after the hurricane struck. About 3 percent of the drinking water facilities in Mississippi were out of service. Wastewater facilities in Louisiana bore the brunt of damage, with 29 percent becoming inoperable. Damage to utilities from this storm was estimated at more than $2 billion.

## Case Study

### Effects from Hurricane Katrina: New Orleans Sewerage and Water Board

The effects from Hurricane Katrina in New Orleans have been long-lasting. As the center of Hurricane Katrina passed southeast of New Orleans on August 29, 2005, winds downtown were in the category 3 range, with frequent intense gusts and tidal surge. Hurricane-force winds were experienced throughout the city, although the most severe portion of Hurricane Katrina missed the city. The hurricane made final landfall in eastern St. Tammany Parish. The western eye wall passed directly over St. Tammany Parish as a category 3 hurricane on August 29, 2005. The storm surge in the region was as high as 16 ft (4.9 m). In New Orleans, the storm surge caused more than 50 breaches in drainage canal levees and in navigational canal levees, and precipitated the worst engineering disaster in US history. By August 31, 2005, 80 percent of New Orleans was flooded, with some parts under 15 ft (4.6 m) of water. Most of the city's levees, designed and built by the US Army Corps of Engineers (USACE), broke. These breaches were responsible for most of the flooding, according to a June 2007 report by the American Society of Civil Engineers.

The New Orleans Sewerage and Water Board (NOSWB) had fulfilled its primary duty of protecting the city from the effects of a massive tropical storm system. However, it was soon evident that water was rising throughout the East Bank of the city, although there was no appreciable flooding suffered to this point from the hurricane itself. The source of the water was numerous breeches in the flood-protection levee system in the 17th Street Canal, London Avenue Canal, and the Inner Harbor Navigation Canal, which apparently were unable to withstand the water levels, which exceeded design criteria, and simply folded in against the weight of the water. NOSWB suffered massive losses in water, sewage, and drainage facilities, equipment, supplies, fleet vehicle, and heavy equipment. The result of the flooding on the East Bank of the city was loss of virtually all pumping equipment contained within seven major drainage stations and serious flooding damage to six other drainage stations, plus loss of most of the sewage lift and pumping stations, including the entire East Bank Sewage Treatment Plant. Of even greater impact was flooding of the power house and boiler room and the complete loss of a 25-Hz generating plant; this had never occurred in the history of the board.

The first step in recovery of the city was to restore the power plant and water purification to operating conditions. As a result of the flooding that followed the hurricane, the utility's 61-MW generating capacity had been reduced to 6 MW by loss of three generators. The water plant and support services were placed back on line using a single 6-MW generator after the Herculean effort of a handful of employees and on-the-spot invention of a workable method referred to as "black startup."

Restoration began in earnest on Saturday, September 3, when the handful of NOSWB engineering, maintenance, and operating personnel were joined by the General Electric Disaster Management Team. One result of this partnership was the installation of a new switchgear for the number 3 generator in one week as opposed to the standard 26 weeks. To facilitate restoration, two temporary workforce housing units were set up by General Electric, one in LaPlace to house and feed up to 200 journeyman craftsmen and one to house homeless board employees and the General Electric emergency management teams. Among the improvements installed after Hurricane Katrina was a $3.5 million, 15-MW generator installed on the grounds of the Carrollton Water Purification Plant, which will give the board more reliable power for drainage and water operations in the event of a commercial power failure. This was funded 100 percent by USACE. NOSWB crews repaired more than 80,000 water main breaks, house service leaks, broken water meters, and fire hydrant leaks. Full repair of all damage done to the system will take another 15 years; this includes waterline replacement and sewer line replacement, along with associated road repairs. To date, the Federal Emergency Management Agency (FEMA) has allocated nearly $500 million in public assistance funds. In addition, USACE is providing $204 million for restoration. Finally, FEMA has also provided $161 million in hazard mitigation planning grant funds for flood-proofing improvements.

*Source:* Bob Moeinian, PE, Retired Chief of Operations, New Orleans Sewerage and Water Board.

**Floods.** Few areas can be considered completely safe from flooding. The USACE estimates that annual flood damage in the United States amounts to hundreds of millions of dollars, with some events causing damage in billions.

Flooding can be caused by heavy rainfall, snowmelt, hurricane storm surges, high tides, volcanic eruptions, and dam or levee breaks. A flood can contaminate water supplies, inundate transmission and treatment facilities and pumping stations, scatter and ruin stockpiled equipment, and disrupt power, transportation, and communications. Floods from precipitation are relatively easy to predict and mitigate, provided sufficient funds can be obtained to take preventive measures. However, floods can also occur unexpectedly as a result of a dam break, breaching of underground tunnels, or unusually heavy rainfall. For example, the Des Moines, Iowa, water treatment plant was flooded and taken out of service during the summer of 1993 when the Raccoon River overflowed its banks. Flood elevations at the Des Moines plant were approximately 8 ft above the 100-year flood elevation. Water service to 250,000 people was cut off for several days.

More recently, flooding of the Cumberland River in May 2010 left many of the Nashville, Tennessee, utilities severely damaged. Government officials considered this event to be a 1,000-year flood event. Two-day rain totals in some areas were more than 19 in. (0.5 m). One of two water treatment plants and both wastewater treatment plants in the Nashville Metro Water Services system were out of service.

---

## Case Study

### Nashville Flooding

Metro Water Services (MWS) is a water, wastewater, and storm water utility. Nashville's water supply comes from the Cumberland River and is treated at two water treatment plants: R.L. Lawrence (Omohundro) and K.R. Harrington, each with a rated capacity of 90 mgd. More than 2,900 miles of pressurized water mains, 41 reservoirs, and 57 water-pumping stations provide water to more than 177,000 customers. The average daily treatment is approximately 100 mgd.

On May 1 and May 2, 2010, middle and western Tennessee experienced record (one in 500–1,000 years) flooding when rains totaling 13–18 in. (0.3–0.45 m) fell over a widespread area. Damages exceeded $2 billion in the metropolitan Nashville and Davidson County (metro Nashville) area alone. The heaviest rain occurred in the uncontrolled basins of the Cumberland River, resulting in extensive, devastating flash flooding as well as unprecedented flooding along the Cumberland River and its tributaries. There were 11 fatalities in Nashville directly attributed to this event, and more than 10,000 homes were impacted by flood waters. Also, 160 MWS staff members (21 percent) were personally affected by the flood, reducing the available work force during the emergency.

On Saturday, May 1, 2010, the K.R. Harrington Water Treatment Plant was flooded as the river rose to a record 51.86 ft (15.8 m), approximately 30 percent above flood stage at this location (Figure 2-6). As the facility began to flood, the staff planned and conducted a controlled shutdown to protect the infrastructure.

Figure 2-6 Flooded K.R. Harrington Water Treatment Plant in Nashville

*Source:* Sonia Allman, Nashville Public Information Officer.

**Figure 2-7 Flooding at the Nashville Omohundro Water Treatment Plant**

Additionally, modeling predicted the Omohundro Water Treatment Plant flooding to be 52 ft (15.8 m) (Figure 2-7). The projected elevation of the Cumberland River was 52.5 ft (16.0 m), which would result in loss of the remaining water treatment plant in Nashville. Sandbagging and pumping water from within the pumping station protected the Omohundro plant from the flood waters, which peaked at 51.86 ft on May 3, 2010. Fortunately, MWS was able to maintain water quality throughout the flood event and continued to meet all state and federal regulations and guidelines.

When the K.R. Harrington plant was taken off line, the utility asked customers to conserve water. Unfortunately, the original reaction was to hoard water, as customers began filling bathtubs, buckets, and other containers because they feared complete loss of water service. As finished water storage reservoir levels dropped to dangerously low levels, the utility again asked customers to conserve water. The Tennessee Department of Environment and Conservation issued a mandatory water conservation order for Davidson and Williamson counties on May 4, 2010. This assisted MWS in providing ample, safe water supply to customers with only one water treatment plant in operation. The order was lifted on June 1, 2010.

The utility's customers responded admirably. Water conservation became "cool" and expected; those who wasted water were publicly criticized on social media (e.g., Facebook). Lamar Advertising provided free billboard ads regarding water conservation; church bulletins and business signs promoted conservation; and neighborhood groups had contests to see who could take the shortest shower and/or come up with the best water conservation idea. Even the Restaurant Association jumped on board; many local restaurants began using paper plates and utensils in addition to only serving water when requested.

Once the floodwaters receded, the risk of waterborne illnesses remained until municipal and home water systems could be fully disinfected. Most infections came from drinking water contaminated with fecal matter. Further, flooding can cause fuels to mix with floodwaters, which makes the sterilization of drinking water more difficult as it becomes harder to boil.

Following are the lessons learned from these events and advice for future event preparedness:

- Understand what other agencies and entities have in place and work closely with them before, during, and after an event.
- Prepare and deliver effective messages to local officials and customers to help curb water demand and make it possible to maintain supplies even while a plant is out of service (even customers reinforced the need to conserve via Facebook).
- Have workable piping interconnections (and relationships) with neighboring utilities for backup emergency water supply.
- Work closely with staff and contractors to monitor and address any potential main breaks during a plant outage to minimize water loss and intrusion potential.

- Ensure that emergency plans include getting key staff on site, even if by boat.
- Appreciate utility staff who will work tirelessly due to "ownership" and pride in operations, and make sure staff take time to rest and care for themselves.

---

**Wild fires.** Forest and brush fires can damage watersheds and water system structures and strain water supplies if used for firefighting. Forest and brush fires can occur anywhere, particularly in areas affected by dry weather or drought. Firestorms in urban and suburban areas are likely to increase in frequency because of widespread development into formerly nonurban forested areas, termed the "wildland–urban interface." Wildfires scorch soils and create ash and debris that can clog rivers and reservoirs, increasing the cost of water treatment for years to come. When trees and underbrush are destroyed, there is less groundcover to absorb rainfall and prevent soil erosion. Storms flush silt and other debris from the fires into rivers, reservoirs, and ultimately into municipal water treatment facilities, adversely impacting the treatment process. This condition has been experienced repeatedly as a result of forest fires throughout the West.

## Dependency and Proximity Hazards

Dependency hazards are the risks that result from supply chain disruptions. The supply chain in a utility is the network of interconnected businesses that are involved in moving the water from its source to the customer's tap. Examples of dependency hazards include the following:

- disruption of services provided by other utilities (e.g., electric power, gas, communication),
- disruption of services provided by key suppliers (e.g., chemicals, fuel, parts and fittings, contractors),
- inability or unwillingness of key employees to come to work,
- inability of key customers to accept production, and
- systems that send water into the facility or transport it out become inoperable.

Proximity hazards cause collateral damages to a water utility when incidents occur close to the utility's assets. Damage can be caused by humans, natural events, and accidents. It is important that the water utility identify all high-risk operations and businesses within its service area and in the region and understand how their failure can impact water service.

An example of this type of threat is a hazardous material release such as the one experienced on the Elk River in Charleston, West Virginia, in January 2014. Spills originate from pipelines, ships, airplanes, motor vehicles, railroad cars, and fixed containers. For example, in 1981, a crop-dusting plane carrying the herbicide 2,4-D crashed into a central California river not far from the intake for a city water supply. Liquid, solid, and gaseous materials can be spilled, and spills can occur at any time and result in injury, loss of life, property damage, and considerable costs to restore normal water utility activity.

Perhaps the most serious impact of a hazardous material spill on a water utility is contamination of surface water and/or groundwater supplies. Emergency response measures should be taken to prevent spilled materials from reaching the water system or to minimize the spread of contamination throughout the system. Damage to the water system and negative effects on its users can be very expensive, not only in terms of cleanup costs but in replacement of system components and disruption to the community.

## Malevolent Threats

Malevolent threats are deliberate acts carried out by person(s) with the intention of causing harm to others and of causing physical destruction. Such acts can be carried out by an insider (e.g., utility employee), outsider (e.g., vandals, criminals, and terrorists), or by collusion between the two. ANSI/AWWA J100 provides suggestions on how to deal with specific malevolent threat scenarios, including attacks with explosives (e.g., in the air, on land, and in the water), various sizes of attack (e.g., small, medium, large, and extra-large), and attacks not involving explosives (e.g., contamination, theft, and cyberattacks).

Malevolent acts include the following:

- *Contamination:* An adversary gains access to and adds potentially toxic chemical, biological, radiological, or pathogenic contaminants to a treatment process, distribution system, storage tank, or treatment chemicals with the intent to contaminate the finished drinking water.
- *Sabotage:* An adversary damages, disables, or destroys equipment and process control systems, causing harm to a utility and its customers. Sabotage can be caused by physical means or through cyberattack.
- *Theft or diversion*: An adversary steals or diverts information, dangerous substances, valuable resources, or similar materials to compromise the safety and reliability of a drinking water system. Theft and diversion can be caused by physical means or through cyberattack.
- *Direct attack*: An adversary makes a direct attack on a critical utility facility using a boat, plane, or automotive vehicle as a weapon or with an assault team comprising one or more assailants. The perpetrators of such malevolent acts are typically well-trained terrorist groups with significant resources and capabilities.

**Cybersecurity.** Department of Homeland Security and intelligence communities agree that cybersecurity is a top concern, particularly as it applies to critical infrastructure sectors including water. Federal officials underscore that the nation faces significant risks of cyber intrusion/cyberattack on process control systems that stem from vulnerabilities in enterprise/business systems and/or process control systems. Vulnerabilities can be focused on employee records (enterprise system), customer records (enterprise system), and utility records and processes that include financial assets (enterprise system) and operations (process control system/supervisory control and data acquisition).

Multiple high-profile cyberattacks on large companies (e.g., Home Depot, Target, Sony, JP Morgan, Yahoo) have served to reinforce the reality that cyber vulnerabilities can impact critical infrastructure systems. As a result, federal actions regarding cybersecurity do have implications for the water sector.

Executive Order 13636, Improving Critical Infrastructure Cybersecurity, directed the National Institute of Standards and Technology (NIST) to work with stakeholders to develop a voluntary cybersecurity framework for reducing cyber risks. AWWA's guidance document, *Process Control System Security Guidance for the Water Sector,* was developed to provide utility managers with an overview of 12 core practice areas that support a consistent and repeatable recommended course of action to reduce vulnerabilities of cyberattacks. A Use-Case Tool was prepared to support voluntary use of the guidance in alignment with the NIST Cybersecurity Framework. The Use-Case Tool generates a prioritized list of recommended controls based on how a utility may apply various process control technologies. For more information, see the AWWA website.

---

## Case Study

### Stuxnet, The First Cyber Threat Able to Reprogram Industrial Control Systems

In July 2010, malicious software code that targets industrial control systems, such as the supervisory control and data acquisition systems used at utilities to control the flow and treatment of water and wastewater, was discovered. This code, dubbed Stuxnet, exploits Windows vulnerabilities and infects its intended target with a malicious computer worm. It is the first known work to target and can reprogram industrial control systems. Stuxnet is highly complex and implements state-of-the-art techniques and capabilities to infect a system, prevent detection, exfiltrate data, and inhibit analysis of the malicious code if it is detected. Given the emergency of this and other new breeds of cyber threats, it is essential that utilities develop and maintain cybersecurity programs to protect their control systems.

*Source:* Edwards (2010)

## Case Study

### Hacker Holds Water Utility Hostage

Two reports from small systems in Missouri indicate that the utility computer systems and password-restricted files had been hacked and held hostage until a ransom was paid by the system. Payment was to be in bitcoin; in both cases, it was approximately $500. In both instances, the systems "thought" they were doing adequate backup; however, flaws were discovered in their procedures. In both cases, information technology experts advised the systems to either deem their files as "lost" or unrecoverable or to pay the ransom because the cost of attempting to circumvent the event would cost more than the ransom. Small systems with little computer expertise offer "low hanging fruit" to hackers.

## Case Study

### Impact of the 2003 Northeast Power Outage on Water Supply

On August 14, 2003, a series of equipment and alarm faults caused a cascade of failures throughout southeastern Canada and eight northeastern states. All told, 50 million people lost power for up to two days in the biggest blackout in North American history. The event contributed to at least 11 deaths and cost an estimated $6 billion. The lack of power significantly affected the water sector in the affected area. Many water systems lost water pressure, which resulted in potential contamination and response actions to protect the public. For example, 4 million customers in eight counties surrounding the Detroit area were under a boil-water advisory until August 18, four days after the initial outage. In Macomb, Michigan, town officials ordered all 2,300 restaurants closed until they were decontaminated after the advisory was lifted. Cleveland also lost water pressure and instituted a boil-water advisory. Many wastewater systems spilled sewage into waterways. A dependency hazard in the water sector was strikingly revealed by this event. Experts indicate that although significant steps to improve the US power grid have been taken, currently there is no way to guarantee the extinction of large blackouts. The frequency of large-scale blackouts remains relatively high.

*Source:* Minkel (2008)

# ACCEPTABLE LEVELS OF SERVICE FOR A SYSTEM

An important step in risk assessment is to understand the consequences if a threat or hazard materializes. To support this task, a utility should develop specific goals and acceptable levels of service under disaster and recovery conditions. An example goal for a utility is to provide quality water while protecting public health, safety, and the environment, including the preservation of watersheds. Meeting this goal should be a continuous process before, during, and after a disaster.

When considering water system service priorities, an obvious approach is to examine the interdependencies that exist with other critical infrastructure and key resource sectors. Figure 2-8 illustrates some of the dependencies that exist between water and other sectors.

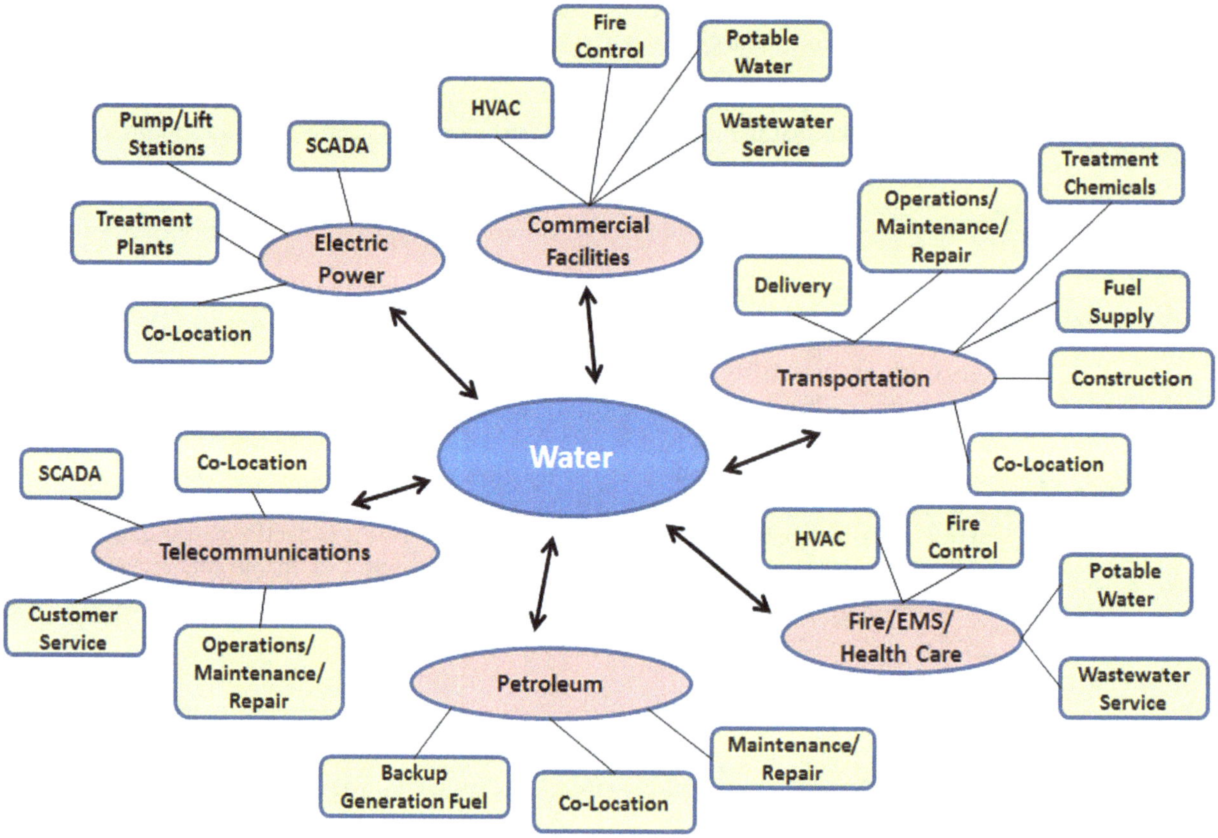


EMS: emergency, management system, HVAC: heating, ventilating, and air conditioning, SCADA: supervisory control and data acquisition
*Source:* Morley (2012)

**Figure 2-8 Water utility interdependencies with other infrastructure systems**

### Establishing Acceptable Levels of Service

Utility requirements during disasters can be estimated based on user needs and the system's capabilities. For planning purposes, a utility should work with local emergency management directors, local hospitals and health care centers, business organizations and large employers, and other priority users to learn more about their service needs. This type of planning is as vital to a utility as the work to strengthen resiliency and mitigate risk.

## CONCLUSIONS

This chapter provides information on hazards and threats, why updating risk assessments is an important activity that enhances a utility's resilience, how often risk assessments should be updated, the standard methodology for initiating or updating a risk assessment, and case studies that show how water utilities have been impacted by natural hazards, human-caused threats, and technological threats. No utility is immune to the sudden and/or disruptive incidents known as "emergencies" that occur from time to time over the life of a utility. However, by understanding how utilities might be impacted by these events through the risk assessment process, managers are in a better place to address whatever the future may bring.

## REFERENCES AND SUPPLEMENTAL READINGS

AWWA. 2011. M60, *Drought Preparedness and Response.* Denver, CO: AWWA.

AWWA. 2010. ANSI/AWWA J100-10. *Risk Analysis and Management for Critical Asset Protection (RAMCAP) Standard for Risk and Resilience Management of Water and Wastewater Systems.* Denver, CO: AWWA.

AWWA. 1989. M32, *Distribution Network Analysis.* Denver, CO: AWWA.

AWWA. 1983. M2, *Automation and Instrumentation.* Denver, CO: AWWA.

AWWA. *Cybersecurity Guidance & Tool.* www.awwa.org/cybersecurity.

AWWA. *Process Control System Security Guidance for the Water Sector.* www.awwa.org/Portals/0/files/legreg/documents/AWWACybersecurityguide.pdf.

Buehrer, J. 1989. Charleston Utility Weathers Hugo's Devastation. *AWWA MainStream,* 33:11:1.

Edwards, D. 2010. Robust ICSs Critical for Guarding Against Cyber Threats. *Journal AWWA,* 102:11:30–33.

Greenemeier, L. 2011. Half-Life and Death: Radioactive Drinking Water Scare in Japan Subsides, but Questions Remain. www.scientificamerican.com/article/fukushima-water-fallout/.

Kennedy/Jenks Consultants Engineers and Scientists. 1993. *Guide Document Earthquake Vulnerability of Water Systems. Prepared for US Geological Survey.* Washington, D.C.: USGS.

Lambe, T.W. and Whitman, R.V. 1969. *Soil Mechanics.* New York, N.Y.: Wiley.

Minkel, J.R. 2008. *The 2003 Northeast Blackout—Five Years Later.* www.scientificamerican.com/article/2003-blackout-five-years-later/.

Morley, K.M. 2012. *Aging Infrastructure: Water Sector Perspective.* Presented at the 2012 Critical Infrastructure Protection Advisory Committee Plenary, Washington, D.C.

Morley, K.M. 2010. Advancing the Culture of Security and Preparedness in the Water Sector. *Journal AWWA,* 102:6:34–37.

Moyer, J.; Dakin, R.; Newman, R.; Groves, D. 2009. The Case for Cyber Security in the Water Sector. *Journal AWWA,* 101:12:30–32.

Tokyo Electric Power Company, Inc. 2012. Fukushima Nuclear Accident Analysis Report (Summary). June 20, 2012.

US Department of Homeland Security and USEPA (US Environmental Protection Agency). 2007. Water: Critical Infrastructure and Key Resources Sector-Specific Plan as Input to the National Infrastructure Protection Plan, May 2007.

USEPA. 2016. Drought Response and Recovery Guide for Water Utilities. www.epa.gov/waterutilityresponse/drought-response-and-recovery-guide-water-utilities.

This page intentionally blank.

# Chapter 3

# Developing an Emergency Response Plan

> *ANSI/AWWA G440, Sec. 4.5.1: The utility shall develop and document preparedness plans in a format conducive with their operations. Preparedness plans must include an Emergency Response Plan and may also include other supplemental plans such as a Resource Typing Plan, Crisis Communications Plan, Disaster Recovery Plan, Mutual Aid and Assistance Plan, and/or a Business Continuity/Continuity of Operations Plan.*

This chapter focuses on development of an emergency response plan (ERP), which is useful in organizing a utility to prepare for, respond to, and recover from an emergency incident. The ERP provides guidance for utility staff during emergencies. This chapter provides proven methods and national standards used to develop and maintain an ERP. This plan establishes a proven management structure, the incident command system (ICS), as the foundation for managing response activities. It also provides utility staff with well-defined response procedures that are intended to restore and/or maintain service delivery during times of crisis. The ERP is designed to address probable system failures. In all response activities, whether foreseen or unforeseen, safety, good judgment, and common sense must take precedence.

Utility resiliency can be best achieved when an ERP is part of an overarching program that includes risk assessment, business continuity planning, training, and exercises. In this chapter, preparedness plan types and components are described, with more detail provided for the ERP. A case study from a utility that has activated their plan during their response to incidents is also included. An example ERP framework with brief descriptions of each section is provided in the following.

By the end of this chapter, the reader should be able to:

- use the results of their risk and resilience assessment to build a comprehensive ERP that fits the utility's needs;
- understand the value of developing and updating an ERP from operational, fiscal, legal, and legislative perspectives;
- develop or revise an ERP for a utility using the framework in Appendix A; and
- choose the components of an ERP that are appropriate for use at a utility.

## IMPORTANCE OF A UTILITY PREPAREDNESS PROGRAM

Preparedness planning and exercises can make even a major emergency more manageable and can expedite recovery efforts. However, plan development and exercises require that a utility invest fiscal and human capital. The return on this investment is realized through the organization's resiliency.

## OVERVIEW OF TYPES OF PREPAREDNESS PLANS

Many utilities have several types of preparedness plans, and using the plans together can be challenging without an understanding of how they fit together. Figure 3-1 shows how numerous plans can be combined. Table 3-1 describes the types of plans described in Figure 3-1.

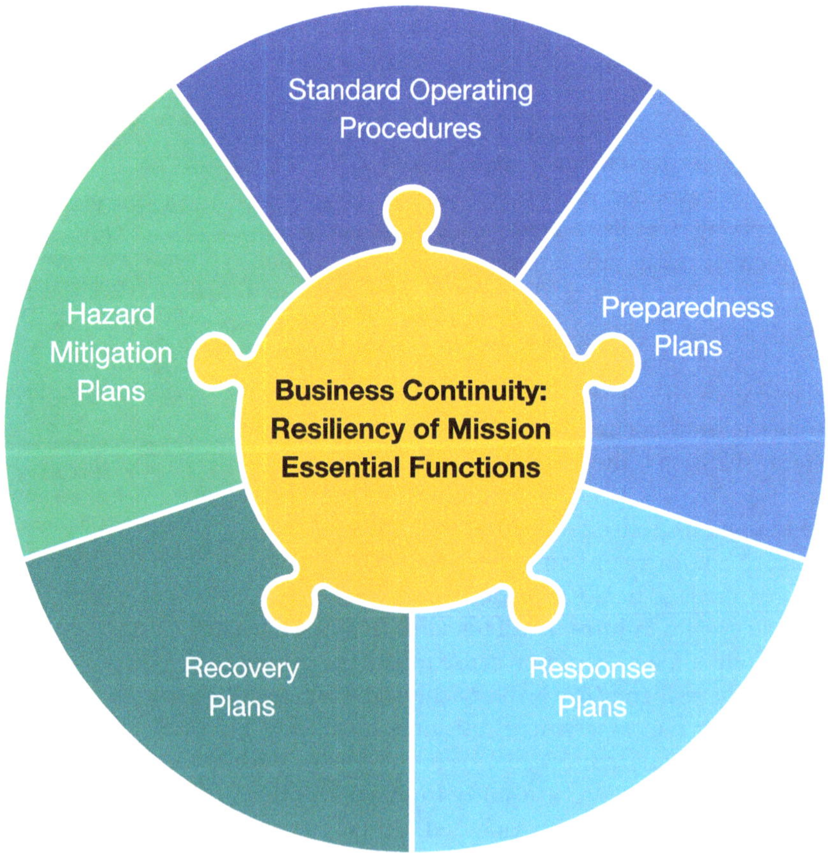


Figure 3-1 Interconnection of various preparedness plans

**Table 3-1 Types of preparedness plans**

| Type of Plan | Included Documents |
|---|---|
| Hazard mitigation | Vulnerability assessment/risk analysis<br>Business impact analysis<br>Impact assessment<br>Hazard identification<br>Risk mitigation plan |
| Standard operating procedures or guidelines to support mission-essential functions | Documentation policy<br>Timekeeping procedures<br>Safety and security plan |
| Preparedness | Facility response plan<br>Hazard-specific (e.g., hurricane, flood) preparedness plan<br>Spill prevention control and countermeasure |
| Response | Emergency response plan<br>Emergency operations plan<br>Fire evacuation plan<br>Incident management plan<br>Crisis communications plan<br>Point of dispensing plan<br>Hazard-specific (e.g., flood, tornado, bomb threat, water quality emergency) response plan |
| Recovery plans | Continuity of operations plan<br>Laboratory continuity of operations plan<br>Pandemic plan<br>Information technology disaster recovery plan |

A business continuity plan (BCP), sometimes referred to as a *continuity of operations plan* (COOP), helps a utility to continue essential operations throughout any event that disrupts business. The BCP is designed to keep the utility's "heart beating" while the ERP is designed to "stop the bleeding" after disaster strikes. For more information on business continuity planning, refer to the Water Research Foundation (2013).

Note that a disaster recovery plan, or "drip plan," is typically the name of the information technology (IT) recovery plan. A utility's IT recovery plan can be a component of the BCP.

## HOW TO DEVELOP AN ERP

After the events of September 11, 2001, all US utilities that serve more than 3,300 people were required, through the Bioterrorism Act of 2002, to conduct a hazard and vulnerability assessment and to develop an ERP. ANSI/AWWA G440, Emergency Preparedness Practices, was developed as a reference standard for utilities to comply with the act. The following basic principles should be recognized when developing an ERP, which should reference existing resources, be concise and logical, and be coordinated with other partner agencies.

### Steps to Developing an ERP

Utility staff should follow the steps described below when developing a useful ERP.

1. *Establish a team to develop and review the ERP.* The team members will be based on the utility's size and complexity. Team members are subject matter experts in utility management, emergency management, water and wastewater treatment

and operations, supervisory control and data acquisition operations, human resources management, and warehouse procurement. Other members include representatives from police, fire, emergency management, and the local emergency planning committee. Those with a role in execution of the plan should also be on the development team.

2. *Review the results of the risk assessment and other available materials (e.g., after-action reports [AARs]).* The risk assessment (see Chapter 2) details the methodology for identifying a prioritized list of threats that the utility faces. AARs (see Chapter 6) from both real emergencies and exercises detail lessons learned and improvement activities. The information from these documents is vital for the team to effectively manage the emergency preparedness program.

   The ERP should reference only existing resources. The plan should not be based on mitigation actions or system improvements that are not in place. For example, plans should not list a portable generator as a standby power source if the generator has been budgeted but not purchased. The ERP should not be revised until the generator is installed, tested, and fully integrated into the system. Subsequent improvements will prompt additional plan revisions and refinement.

3. *Schedule and hold workshops with the team.* Keeping the momentum moving to complete the ERP involves proper project scheduling. A workshop schedule should be established with the team at the start of the project. These workshops will become the stepping stones to a complete document. Some utilities prefer that these meetings last no more than 2 hours, and others hold half-day or full-day workshops.

4. *Coordinate with other partner agencies (e.g., federal, state, provincial, local, or others, depending on country and state of origin).* An effective ERP requires the cooperation of many agencies and other lifeline services. Utilities depend on electric, transportation, and communications networks to function.

   Public and private organizations may need to work together on a community level to prepare for, respond to, and recover from a disaster. Coordination during development of the ERP may assist in the resiliency of all agencies involved.

5. *Write/revise the ERP.* Whether a utility builds a new plan or has an existing ERP to revise, the ERP should be as concise and logical as possible. The wording should be straightforward, and terminology should be familiar to personnel who will implement the plan. The information in the plan should be presented in a logical order, with the most important information, such as the communication chart and emergency telephone numbers, readily accessible.

6. *Adopt and roll out the ERP to staff and provide training.* Formal adoption of the ERP by the utility is important to ensure the most recent version is applied. Record-keeping and simple version control documentation inside the front section of the ERP can help over the long term. Utility leaders should officially distribute the document to all staff who will use it (this can be done electronically and/or by paper), with a note to staff about the document's importance and security. Training sessions for all staff should follow within one month of plan distribution (for more information on training, see Chapter 6).

7. *Revise and update the ERP at least yearly and after each exercise or event.* An out-of-date ERP is not very helpful to utility staff. Procedures evolve and typically change after staff gain the experience of an actual

> ***ANSI/AWWA G440, Sec 4.5.2: Maintain the Plans.***
> *The utility shall maintain the preparedness plan as and when changes are perceived or, at a minimum, on an annual basis.*

emergency or training event. A plan update should be included in the utility master schedule each year, and the ERP should be updated after any actual emergency or training event. The Homeland Security Exercise and Evaluation Program (HSEEP) process (explained in Chapter 6) uses an AAR. Updates should also include information from risk assessment reports, AARs, and mitigation plans.

## COMPONENTS OF AN ERP

An ERP is a working document that should be used before, during, and after a disaster. The plan is the logical outgrowth of, and developed from, the hazard summary, risk assessment, and implemented mitigation actions. The following sections describe basic information that should be contained in a plan. Although the elements are presented in a recommended order, individual systems may modify, add, or delete sections as necessary. The contents of a plan may be dictated by state, provincial, or local regulations or guidelines.

Utilities may expand or contract these sections based on their size and individual utility needs. Guidance for building or revising an ERP can be found in a guide from the Federal Emergency Management Agency (FEMA 2010) and in Appendix B. Note that these are two of many different styles and approaches to preparing an ERP; the plan contents are flexible based on the utility's needs.

### ERP Content

Emergency procedures are an important part of any emergency preparedness program. The ERP is intended to provide a comprehensive overview of all emergency preparedness program elements. Much of the plan may be in narrative form, explaining the rationale and basis for the emergency preparedness program; it is not designed for use during response activities. Procedures, on the other hand, are specifically designed for use during an emergency and to provide step-by-step guidance for implementing critical aspects of the emergency preparedness program. Procedures may be organized in a checklist format, with a completion block provided for each task to facilitate the implementation process and to create a record of task completion.

Generally, there are two types of procedures: positional and functional. Positional procedures are created to provide guidance for each position within the organization. In most instances, they detail the specific duties and responsibilities of each position, as well as detailed tasks or actions needed for various emergency situations or conditions. If the utility uses the ICS, as recommended in ANSI/AWWA G440 and Homeland Security Presidential Directive 5, procedures are developed for each position within the utility ICS. If not, the procedures are developed for each position in the emergency organization established by the utility. Each procedure should include a listing of additional resources that will be needed to fully implement the responsibilities of the position. This might include call lists, emergency resource lists, and functional procedures.

Functional procedures are developed to provide detailed step-by-step guidance for implementing a particular function of the emergency preparedness plan. Functional procedures may be developed for such activities as implementing emergency notifications, performing emergency shutdowns, and developing emergency interconnects. The purpose of a functional procedure is to guide the user through the implementation of an emergency function. Such functions are typically implemented infrequently, and the procedure can ensure the emergency function is done quickly and correctly under emergency conditions. Procedures are particularly important for activities that involve multiple tasks that should be performed in a particular order.

Critical components of an ERP include the following:

- System overview;
- Incident management overview;
- Roles and responsibilities, based on the National Incident Management System (NIMS) and ICS (see Incorporating the NIMS and the ICS into an ERP section);
- Internal and external contact information, which should be consolidated in selected locations in the ERP or its appendices to make updating the contact information as easy as possible;
- General emergency response guidance;
- Communications plan (see Chapter 5); and
- Record-keeping.

In addition to general emergency response guidance, the ERP should include plans for the unique aspects of specific hazards. Example incidents for hazard-specific plans include the following:

- main break,
- tank failure,
- pump failure,
- major supply source failure,
- treatment system failure or malfunction,
- waterborne disease outbreak,
- potential or confirmed contamination event,
- loss of positive pressure,
- chemical spill or release,
- chemical supply shortage,
- power outage,
- facility fire,
- flooding,
- wildfire,
- earthquake,
- volcanic eruption,
- severe weather,
- bomb threat, and
- civil unrest.

Any hazards identified to be priorities in the risk assessment should be addressed in the ERP. In the appendix of the ERP, procedural checklists that define actions to take during specific events should be included. These can be used as "rip-and-run" sheets to pull from the book and notate during an emergency. If these procedures are in an appendix, they can be easily modified and updated without reapproval of the entire ERP. Sometimes these response protocols are detailed enough to warrant a separate planning document and are then referred to as hazard-specific plans or grab-and-go binders.

Other content that is often included in the ERP appendices includes the following:

- ERP distribution list,
- facility points of contact,

- emergency response personnel,
- external first responder and government agencies contact list,
- contact list of available emergency contractor services,
- chemical suppliers contact list,
- other vendors contact list,
- contacts for normal and emergency water interconnections,
- emergency operations centers (i.e., local, city, county, state),
- information on the Water/Wastewater Agency Response Network (WARN),
- accredited laboratories contact list,
- power utility and supplier,
- bomb threat checklist,
- US Environmental Protection Agency (USEPA) threat (contamination) identification checklist,
- list of major asset locations,
- schematic maps and aerial photos of major systems and facilities,
- water tank and reservoir levels,
- chemical tank locations,
- emergency response cost tracking form,
- incident log, and
- damage assessment forms.

In addition to appendices, annexes are often used as part of an ERP. Annexes are generally used for other plans that relate to the ERP, such as the BCP or COOP, pandemic plan, and others listed in Table 3-1. Hazard-specific plans are sometimes included as "hazard-specific annexes," and ICS plans are sometimes included as "functional annexes".

Several reference documents are available to help a utility prepare response checklists. The all-hazards consequence management plan was developed jointly by the USEPA and the Preparedness, Emergency Response, and Recovery Critical Infrastructure Partnership Advisory Council Working Group (2009). This plan describes the procedures that a utility can take to minimize the consequences from a variety of incident types. In addition, USEPA's Small and Medium Utility EPR Guidance contains several sample action checklists.

USEPA has also developed 10 rip-and-run style checklists to facilitate emergency preparedness, response, and recovery actions for water utilities. Links to these checklists can be found in Appendix A.

## Planning Considerations

**Agreements with other agencies and organizations.** Many utilities include written agreements with other agencies, utilities, and response organizations within their ERP appendices. These agreements include interconnection agreements with other systems if they are in force. Particularly important are contracts with private suppliers and contractors. Usually, these contracts detail the nature, conditions, and cost of contracted services. Most contracts have an expiration date, so they should be updated periodically. One important administrative tip is to try to get all service contracts on the same renewal schedule and to designate someone to confirm they are updated before expiration. Some utilities require 24-hour access and emergency phone numbers from their suppliers as part of their annual bidding package and contract. All utilities are encouraged to participate in their state WARN (see Chapter 4) and thereby ease the legal agreement process.

**Disaster recovery accounting.** A utility should identify and document all costs related to each emergency it experiences. It may be possible to recover all or a portion of these costs through federal, state, or provincial disaster relief funding. One method to clearly identify the costs related to a disaster is to use a preplanned emergency work order numbering system. All costs related to a specific operation are charged to a unique account number. Many work orders may be generated during one response effort because there may be many different projects in progress to restore service delivery. However, for rigorous cost accounting, all emergency work orders should be specifically coded to identify them as being related to the emergency.

It also helps to use an operating and damage report form to track the work. Such records can be invaluable when assessing the damage of a disaster and evaluating the response. The records also indicate if follow-up actions are needed. Photos and video are excellent ways to document actions. FEMA has several prepared forms for tracking disaster recovery work.

**Distribution list.** A distribution list that identifies the individuals who have copies of the emergency plan should be maintained. The list will allow tracking of all plan updates and modifications.

The Department of Homeland Security (DHS) has identified 16 critical infrastructure sectors that are interdependent in our society. The 16 critical infrastructure sectors as identified in Presidential Policy Directive 21: Critical Infrastructure Security and Resilience are as follows:

- Chemical Sector
- Commercial Facilities Sector
- Communications Sector
- Critical Manufacturing Sector
- Dams Sector
- Defense Industrial Base Sector
- Emergency Services Sector
- Energy Sector
- Financial Services Sector
- Food and Agriculture Sector
- Government Facilities Sector
- Healthcare and Public Health Sector
- Information Technology Sector
- Nuclear Reactors, Materials, and Waste Sector
- Transportation Systems Sector
- Water and Wastewater Systems Sector

A utility should consider discussing their interdependencies in a workshop with representatives from the other independent sectors. They can then design the coordination and procedures in the ERP based on discussion results. Examples of the dependencies in the water sector are requirements for electrical power and treatment process chemicals that are dependent on the electrical grid and transportation. Other sectors are dependent on the water sector, for example, hospitals that require the functions of water and wastewater utilities.

## INCORPORATING THE NIMS AND THE ICS INTO AN ERP

*ANSI/AWWA G440, Sec. 4.5.3: The utility shall be familiar with the concepts of the National Incident Management System (NIMS), to include the Incident Command System (ICS). The utility may include NIMS and ICS in its preparedness plans. Incorporation contributes to an enhanced coordination between utility personnel and outside first responders.*

The NIMS, of which the ICS is a part, provides a systematic, proactive approach to guide utilities and all agencies, whether public or private, to work seamlessly in handling any crisis. NIMS provides the template for management of an incident, while the National Infrastructure Protection Plan provides the structure and mechanisms for a national-level policy related to incident management. ICS provides a common organizational structure for a utility to easily coordinate with other first responders while responding to a crisis. Use of an ICS dates to the Roman times and was more recently noted for its ease in fighting wildfires in California in the 20th century.

Figure 3-2 is a diagram of the ICS structure for a utility to use for their crisis response. Ideally, the utility organizes in this way for common incidents such as a water main break so that the ICS organizational structure and common language become familiar to the staff. The staff will then find it easier to use the same ICS structure during a larger emergency incident.

### Benefits of NIMS Compliance

NIMS compliance is required to receive federal emergency response and preparedness funding. All grant funding for disaster mitigation, preparedness, and response requires compliance. FEMA requires local agencies to "institutionalize" the ICS and NIMS through the following actions:

- **Adoption**
    - Adopting the NIMS through executive order, proclamation, or ordinance that applies to all departments and agencies, as well as promoting and encouraging NIMS adoption by associations, utilities, nongovernmental organizations (NGOs), and private sector emergency management and incident response organizations.

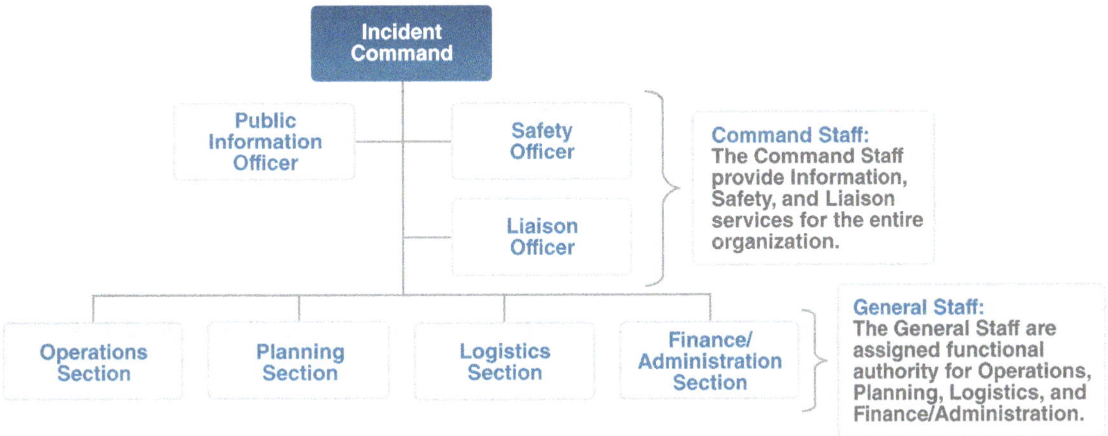


*Source:* FEMA

Figure 3-2 ICS diagram

- Designating and maintaining a single point of contact within the agency to serve as the principal coordinator for NIMS implementation jurisdiction wide (to include a principal coordinator for NIMS implementation within each department/agency).
- Ensuring that federal preparedness awards such as the DHS Grant Program and Urban Area Security Initiative funds support all required NIMS compliance objectives.
- Ensuring that audit agencies and review organizations routinely include NIMS compliance requirements in all audits associated with federal preparedness awards.

- **Planning**
  - Integrating the ICS into functional and system-wide emergency operations, policies, plans, and procedures such as emergency response and operations plans and standard operating procedures.
  - Promoting and/or developing intrastate and interagency mutual aid agreements and assistance agreements to include agreements with the private sector and NGOs.
  - Planning for special-needs populations in the development of ERPs (including individuals with limited English language proficiency, individuals with disabilities, children, the aged, and similar populations).
  - Including preparedness organizations and elected and appointed officials in the development of ERPs.
- **Training and Exercises**
  - Using existing resources such as programs, personnel, and training facilities to coordinate and deliver NIMS training requirements.
  - Conducting ICS training for responders, supervisors, and command-level staff.
  - Planning for and/or participating in an all-hazards exercise program (e.g., HSEEP) that involves emergency management/response personnel from multiple disciplines and/or multiple jurisdictions.
  - Incorporating corrective actions into preparedness and response plans and procedures.
  - Including NGOs and the private sector in an all-hazards exercise program, when appropriate.
  - Promoting the integration of ICS, multiagency coordination systems, and public information into appropriate exercises and comparing against associated target capabilities (refer to HSEEP).

## ERP MAINTENANCE

As the final step to building an ERP, the document should be updated and/or reviewed at least annually and after each emergency or significant changes to the utility that affect the utility's risk profile, emergency preparedness, and response. Contact information should be updated quarterly. The ERP should also be updated whenever operational or policy changes are made that affect response procedures. The responsibility to maintain the document should be assigned to one person at the utility. This will provide focus and the directive to keep the ERP up-to-date.

## Case Study

### Large-Utility Example: Multiagency Response to a Major Water Pipe Break

The Massachusetts Water Resources Authority (MWRA) was established in 1984 to provide wholesale water and sewer services to 2.5 million people and more than 5,500 large industrial users in 61 metropolitan Boston communities. MWRA provides drinking water to residents who live in 51 of these communities. At about 9:30 a.m. on Saturday, May 1, 2010, a 10-foot-diameter coupling broke loose near Shaft 5A of the MetroWest water supply tunnel, a primary component of the MWRA system. Millions of gallons of water were released. The MWRA's actions, which were required to maintain water supply and repair the pipe damage, affected the drinking water supply to the city of Boston and 29 other communities in the metropolitan area. To maintain water delivery, MWRA reconfigured parts of the transmission system and switched to a combination of fully treated and chlorinated backup water supplies. Massachusetts Governor Deval Patrick declared a state of emergency and issued a boil-water order for the affected communities. Despite the magnitude of the situation, MWRA repaired the broken pipe and restored full flow in less than two days. By 7:00 a.m. on Tuesday, May 4, the boil-water order had been lifted for the 30 communities affected by the break. MWRA and its response partners executed the emergency response professionally, efficiently, and effectively. Interviews with members of responding agencies and organizations and media content analysis showed a high level of satisfaction with the response. This case study identified both best practices and challenges.

The case study presented here describes a large utility's effective response to an emergency. The MWRA had developed an ERP and exercised the same scenario four years before it happened. The following three actions taken by the MWRA are important for ensuring an effective emergency response.

- *Develop a preparedness culture*. The Shaft 5A response underscores the importance of thorough emergency planning and preparedness and illustrates the critical need for drinking water and wastewater utilities to engage in mindful and active emergency preparedness programs. MWRA and several of the other responding agencies have well-established preparedness cultures, and this emphasis on emergency response planning and training is a fundamental reason for the success of the Shaft 5A response. This preparedness-focused outlook can be maintained and bolstered through a variety of the practices illustrated in this case study, including the following: conduct emergency response planning; emphasize drills and training; treat normal incidents as drills; learn the ERP but also emphasize the importance of mindful response improvisation; and review and revise the ERP, incident command structure, and associated emergency protocols after every emergency event. See Chapter 1 for more information on developing a preparedness culture.
- *Nurture relationships before an event, not just during an event.* Another critical reason for success illustrated by this case study was that the response was facilitated by substantial cooperation and coordination among a wide range of emergency responders. This focus on collaboration was exhibited both within MWRA and at the interagency level. Colleagues in different agencies and departments knew and trusted each other, had participated in drills together, collaborated with each other during other emergency and nonemergency events, and knew what to expect from one another. Thanks in part to the involvement of the Massachusetts governor and the mayor of Boston, the various state and local agencies moved easily into designated roles and maintained effective communication among all the players. The Shaft 5A event provides a model for other utilities on how to nurture interagency relationships by developing trust, mutual respect, and

knowledge of one another's systems, capabilities, and communication practices. For more information on building relationships with key partners, see Chapter 4.

- *Develop a strong emergency operations center/incident command.* Key response team members involved in the Shaft 5A incident met soon after the break was discovered and implemented a command structure in which the Massachusetts governor was in charge of state operations, the Boston mayor was in charge of the city's operations, MWRA's executive director was in charge of coordinating MWRA's operations with the state, and the MWRA chief operating officer was the incident commander for MWRA. The greater Boston emergency response structure enjoys many positive characteristics, including strong preparedness cultures within agencies, well-trained staff, and excellent interagency relationships. These traits, combined with this effective command structure, enabled the region to deal successfully with the Shaft 5A emergency. Utilities should articulate and implement a coordinated, prearranged command structure for a successful emergency response. This should entail use of an ICS approach such as the FEMA ICS and development of an explicit and robust system of emergency command (including succession of authority and responsibility), control, and communication, including designation of a public information officer who oversees coordinating all external communications.

## REFERENCES AND SUPPLEMENTAL READINGS

American Water Works Association Research Foundation and Japan Water Works Association. 1993. *Instrumentation and Computer Integration of Water Utility Operations*. Denver, CO: AWWA/AwwaRF.

ANSI/AWWA G440. 2017. *Emergency Preparedness Practices*. Denver, CO: AWWA.

Contra Costa County California Health Services Department. 1992. Risk Management and Prevention Program. Martinez, CA: Contra Costa County Health Services Department.

Hill, K. 1992. Disaster Preparedness Case History: Charleston CPW's Water Distribution Department Responds to Hurricane Hugo. Distribution System Symposium Proceedings 1991, pp. 33–39. Denver, CO: AWWA.

Pickett, M.A.; Laverty, G.L.; Abu-Yaesin, O.A.; Lay, C. 1991. Lessons Learned From the Loma Prieta Earthquake. *Journal AWWA*, 83:11:34–39.

Preparedness, Emergency Response, and Recovery Critical Infrastructure Partnership Advisory Council Working Group. 2009. *All-Hazard Consequence Management Planning for the Water Sector*. www.awwa.org/Portals/0/files/resources/water%20knowledge/rc%20emergency%20prep/All-Hazard.pdf.

Ryckman, D.W. and Ryckman, M.D. 1980. Organizing to Cope With Hazardous Materials Spills. *Journal AWWA*, 72:4:196.

Sander, D.K. 1991. Contingency Planning: Do You Know What To Do in Case of an Emergency? AWWA Annual Conference Proceedings 1991, pp. 443–456. Denver, CO: AWWA.

US Department of Homeland Security. 2013. National Infrastructure Protection Plan.

US Federal Emergency Management Agency. 2010. Comprehensive Preparedness Guide: Developing and Maintaining Emergency Operations Plans, version 2.0.

Water Research Foundation. 2013. *Business Continuity Planning for Water Utilities: Guidance Document*. WRF/AWWA/EPA Web Report 4319. www.waterrf.org/resources/pages/PublicProjectPapers-detail.aspx?ItemID=6.

Chapter 4

# Mutual Aid and Partnerships

This chapter describes why partnerships and mutual aid agreements are a vital component of an emergency preparedness program and how they can increase a utility's resilience. These partnerships can include other critical infrastructure operators; first responders; community representatives; local, state, and federal agencies; and other utilities through the Water/Wastewater Agency Response Network (WARN). The slogan for WARN is "utilities helping utilities," and that is exactly what it does.

It is imperative that each utility determine their interdependencies before there is a disaster. By forming partnerships and agreements before there is a need, utility staff can be certain that support has been organized for deployment when an emergency occurs. Once a disaster occurs, there is no time to form partnerships and relations with outside organizations and interdependent partners. The lack of partnering can significantly impede a utility's ability to maintain lifeline services to the community it serves.

> ***ANSI/AWWA G440, Sec. 4.8.1:*** *The utility shall identify key agencies that are essential to emergency response and recovery and establish and maintain collaborative partnerships with these agencies. Customer community agencies, such as police and fire; and managers of critical interdependent infrastructure, such as power companies, first responder agencies, and adjacent utilities, are typically included as key agencies.*

## IDENTIFY KEY PARTNERSHIPS

Utilities should identify key agencies that are essential to their day-to-day operations and during emergency response and recovery operations. The next step is to develop and maintain collaborative partnerships with those agencies. This can be done by hosting a local "brainstorming session" with neighboring utilities and other agency staff to identify the outside agencies to coordinate with for emergency scenarios in the utility emergency response plan.

Following are some of the interdependent agencies with which a utility may need to coordinate agreements or collaborations in their community. Because each community is different, the brainstorming session should allow for a free and wide-ranging discussion.

- Fuel
- Electricity
- Parts for rolling and stationary equipment
- Chemicals
- Pipe and supplies
- Trained/credentialed staff
- Rental agencies
- Pump and motor services
- Big box stores (e.g., Wal-Mart, Target)
- City services
- County services
- State agencies
- Primacy agency (e.g., health department or department of environmental services)
- Police
- Fire
- Adjacent utilities
- Interconnecting utilities
- Transportation
- Lodging/food, water, ice for mutual aid responders
- Universities
- Bottling companies
- Software companies (e.g., Microsoft)
- Laboratories
- Telecommunications

## ESTABLISH COLLABORATIVE PARTNERSHIPS

An official partnership should be established with each key partner identified by the utility. There is an existing relationship with a utility's critical vendors; discussions to create a formalized agreement should be initiated so that the utility retains priority with the vendor for support during an emergency. For example, a utility should ensure that local chemical suppliers can guarantee chemical deliveries from outside a predictable disaster zone in a timely fashion at contract rates.

Many utilities have limited relationships with emergency responders and coordinators. Yet, when

> ***ANSI/AWWA G440, Sec. 4.8.2:*** *The utility shall establish collaborative partnerships with key agencies and other groups as appropriate (e.g., local emergency planning committee, local fire department, local police department, regional and state department of Homeland Security, Water/Wastewater Agency Response Network [WARN], neighboring water utilities), to ensure cooperation and effective coordination during emergency response recovery.*

disaster strikes, these resources are essential to coordination of response and recovery. Before a disaster occurs, utilities should contact their local emergency management groups as appropriate, for example, local fire department, police department, regional Department of Homeland Security (DHS), and city, county, and state emergency management agencies. Contact information for key partners at the state and federal level can be found on the US Environmental Protection Agency (USEPA) website.

Most counties have an emergency management division that is responsible for all emergency declarations within the county. Coordination meetings are usually held quarterly, and a utility representative is typically welcome to attend. Creating "face-to-face" relationships with the attendees at these meetings will help the utility representative gain the coordination needed during an emergency.

## MUTUAL AID

Utilities are committed to helping each other and to taking steps that will encourage other utilities and local and state governments to establish intrastate mutual aid and assistance networks. The purpose of these networks is to provide a method whereby utilities that have sustained damages from natural or human-caused events can obtain emergency assistance in the form of personnel, equipment, materials, and other associated services, as needed, from other utilities. The objective is to provide rapid, short-term deployment of emergency services to restore the affected utility's critical operations.

A preestablished agreement among a network of utilities can complement and enhance state and local capabilities to prepare for and respond to a broad range of threats, both natural and human caused. The establishment of such intrastate mutual aid and assistance networks is a core principle of the National Preparedness Goal and the National Incident Management System, developed by the DHS. Formalization of the existing capability to provide mutual aid and assistance provides utilities with a degree of resiliency during natural and human-caused disasters that ensures the continuity of service delivery.

## MUTUAL AID IN THE WATER SECTOR

The water sector has taken steps to significantly enhance resilience through the establishment of mutual aid and assistance agreements. WARN is an initiative that provides a formal agreement for utility-to-utility mutual aid and assistance within nearly every state and several Canadian provinces (Figure 4-1). Utility peers with specialized expertise and equipment are readily available to provide rapid response when incidents occur. Participation in such agreements is traditionally at no cost and does not obligate signatories to respond.

Several states have instituted statewide mutual aid legislation that authorizes the state and its political subdivisions to provide emergency aid and assistance for all disciplines (e.g., fire, law, medical) in the event of a threatened or actual major disaster. In most cases, a proclamation of an emergency by the local government body or the governor triggers the activation of mutual aid resources. Note that the provisions of these laws do not typically preclude utility participation in a sector-specific mutual aid network. In the case of the water/wastewater network discussed here, a

> *ANSI/AWWA, G440 Sec. 4.8.3: The emergency response and recovery plan should include provision for the utility to participate in mutual aid and assistance agreements at the local, regional, and state levels, such as WARN. Such agreements should be documented in the appendix of the plan.*

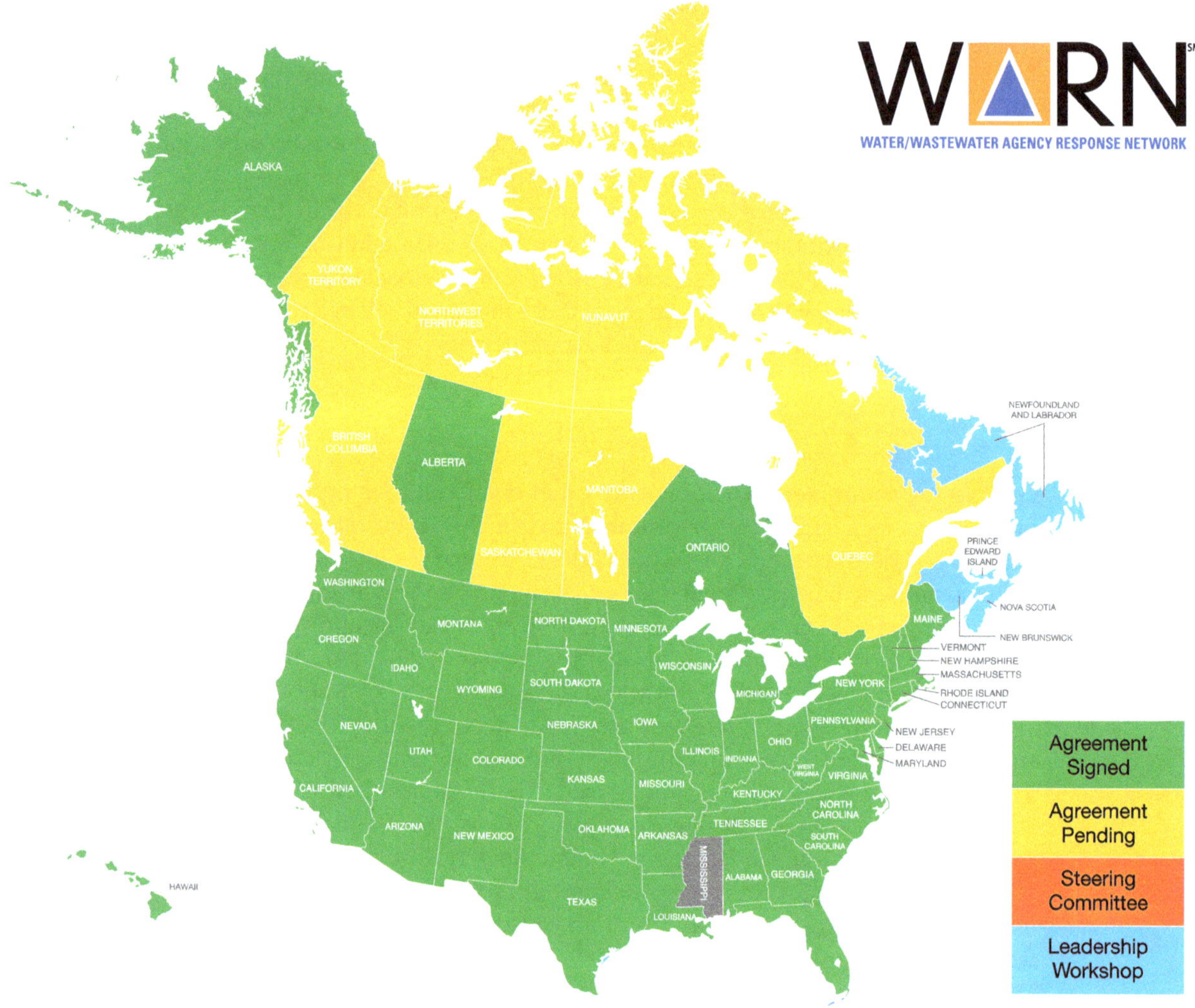


*Source:* National WARN

**Figure 4-1 Water/Wastewater Agency Response Network program map (as of November 2015)**

request for assistance during an incident does not have to be triggered by an emergency proclamation by the local elected body, the governor, or the president. Gubernatorial declarations may allow access to additional aid for public utilities, while the utility mutual aid assistance program may remain independent. WARNs provide a state-level mutual aid and assistance capability through which utilities can request assistance from peers in the absence of a local, state, or national emergency declaration. When a WARN steering committee is formed, this latter point will require discussion and coordination with the state emergency preparedness and response agency.

The ultimate goal for the mutual aid and assistance network program is to create a single intrastate agreement from which all utilities operate and to eliminate multiple statewide utility agreements. If multiple agreements are considered, the determination of which agreement to use could add confusion during a time of critical need and response. Creation of a single comprehensive mutual aid and assistance agreement (MAA) relies on the progression of each of the elements below to put in writing how the program works. The MAA provides the essential framework for how assistance will be provided, outlines response procedures, and addresses issues of insurance and liability.

The member utilities should develop the details of an intrastate utility MAA using the appropriate state law references. The agreement should address all hazards—natural and human caused. According to the DHS, to meet federal grant requirements, MAAs need to account for the following:

- clear definition of key terms,
- roles and responsibilities of individual parties,
- procedures for requesting and providing aid,
- reimbursement and allocation of costs,
- notification procedures,
- protocols for interoperable communications,
- relationships with other jurisdictions,
- workers' compensation,
- liability and immunity,
- recognition of qualifications and certifications, and
- sharing agreements.

MAAs should be in place before an incident to

- help ensure effective response and recovery;
- adequately identify the type and resources needed to provide the specialized resources and employee skills that a utility needs for short-term response and recovery; and
- following a presidential-declared emergency, obtain federal reimbursement as identified in the Stafford Act.

The MAA should include a provision on how it will be updated. Changes to state law, federal regulations, and reimbursement issues may necessitate the need to evaluate and modify the agreement. Keeping the agreement "alive" requires an ability to revise it.

## MUTUAL AID COORDINATION WITH LOCAL AND STATE EMERGENCY MANAGEMENT

As depicted in Figure 4-2, WARN can provide a critical link between utilities and their local and state emergency management agencies, emergency management agency (EMAC), and federal partners. Connection of WARN into local and state emergency operations centers allows multiple agencies to maintain a common operating picture and up-to-date situational awareness regarding water sector impact and mutual aid activities. This connection also ensures that if the WARN is overwhelmed, resources from other mutual aid agreements, state agencies, EMAC, and/or federal resources can be requested to supplement the WARN.

Some WARNs maintain a "distributed" approach in which there is no central coordination and the only contact for WARN resources occurs between the utility that requests assistance and the utility that aids. While this reduces the burden of central coordination, it presents challenges for coordination and information-sharing with local and state EMACs, state primacy agencies, and federal support agencies. In these distributed WARN systems, it may be necessary for the state EMAC or primacy agency to provide central coordination needed to ensure a common operating picture and situational awareness.

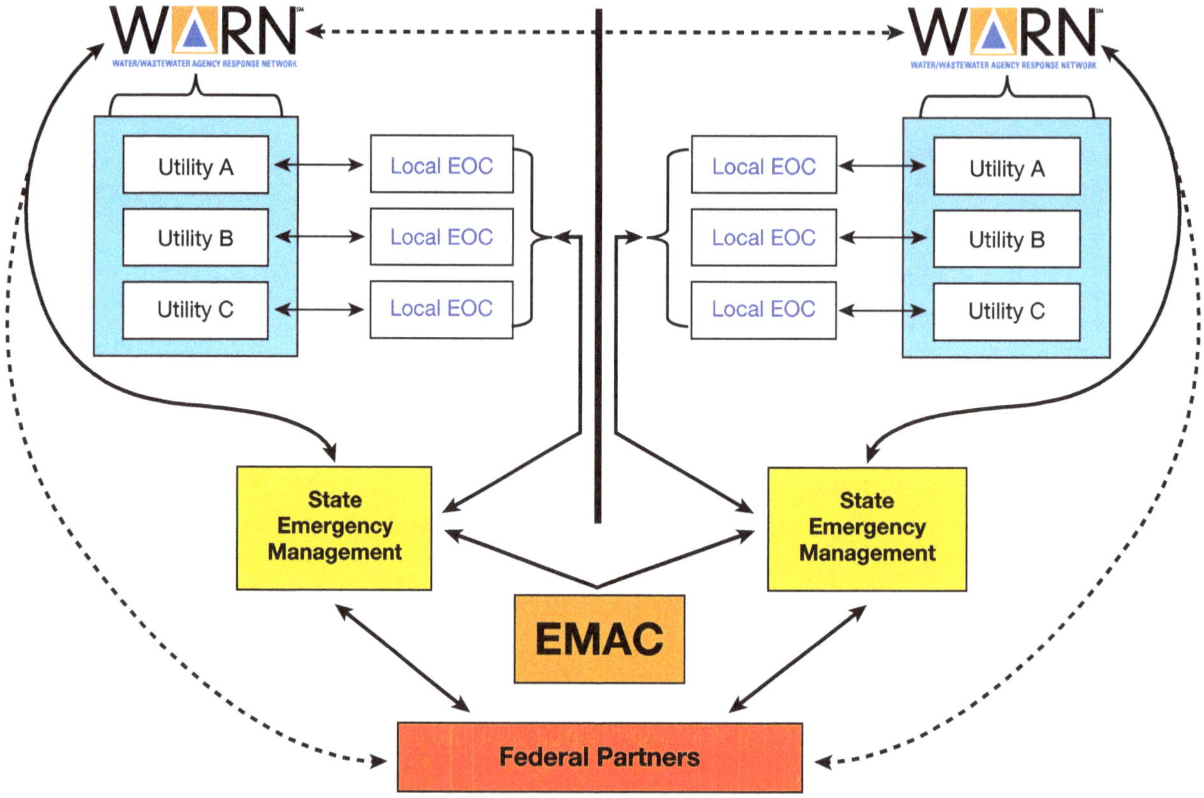


EMAC: emergency management agency, EOC: emergency operations center
*Source:* Morley (2015)

**Figure 4-2** Water/Wastewater Agency Response Network coordination with local, state, and federal emergency management

## WARN OPERATIONAL PLANS

The WARN agreement addresses the legal and administrative functions, and the WARN operational plan outlines the procedures that support the implementation and activation of the WARN agreement during a response. WARN operational plans should not address command and control activities at the field level but should address the coordination elements within the WARN and emergency operations centers. This enables the integration of WARN utilities' actions before, during, and after an incident. As described in USEPA's draft guidance on developing WARN operations plans, the WARN operational plan achieves the following goals (USEPA 2009):

- Describes the pre-emergency governance structure of the WARN program
  - Many WARNs have a steering committee and other groups that help maintain the agreement and communicate and coordinate with participating utilities and coordinating agencies.
- Describes training, exercises, and procedures to update the WARN operational plan
  - It is important to test the agreement before using it during an actual incident. It may also be useful to have an exercise to build on lessons learned from an incident. Training on basic emergency response such as an incident command system (ICS) may be needed for utilities, as well as WARN-specific trainings on how to use the website or use resource typing.

- Provides a general set of procedures for coordinating with associate members and other response partners
  - As discussed in the previous section, this can be a key element to the successful implementation of WARN.
- Provides a general set of procedures for activating the WARN agreement
  - When a utility is affected by an incident and needs support, these procedures will help them identify how to use the WARN and who to call to start getting the support they need.
- Provides a general set of procedures for mobilization of WARN member utility resources
  - The logistics of sending and receiving resources are often overlooked and require planning and coordination in advance of an incident.
- Provides a general set of procedures for internal WARN response coordination
  - For WARNs that are centrally coordinated, this part of the plan identifies the types of information that may be collected and provided to WARN utilities and support agencies.
- Describes documentation and forms for WARN standard reporting formats
  - Standard forms and information templates can help facilitate a more efficient response.
- Describes communications tools for WARN member utilities
  - Most WARNs have a website, and the operational plan describes how the site is used during a response.
- Describes a general set of procedures for writing an after-action report (AAR) and improvement plan
  - Documentation of lessons learned from exercises and incidents is key to improving the WARN for future activations.

## WARN EXERCISES

Many resources exist to help WARNs conduct their own exercises and provide water sector mutual aid objectives to non-WARN-specific exercises. USEPA has published the *WARN Tabletop Exercises Facilitator Guide,* which provides all the materials needed to plan and conduct an exercise. These materials may also be useful for non-WARN-specific exercises, such as those conducted by local and state EMACs. Integrating water sector objectives, in general, and WARN-specific objectives can be very valuable to ensuring coordination with these agencies during a response. WARNs are encouraged to not only hold exercises with participating utilities but to also invite support agencies at the local, state, and federal levels.

Since 2008, USEPA has provided support to WARNs in the form of contractor assistance to conduct and facilitate WARN tabletop exercises. This support has helped ensure that nearly every WARN has conducted an exercise and has an AAR to help document the improvements needed to the WARN based on the exercise experience. Several common activities have been identified during past WARN exercises, including the following (Whitler 2009):

- develop and refine WARN operational plans,
- strengthen the connection to the emergency management community,
- continue incident command system training,
- implement accounting and documentation procedures,

- incorporate resource management and resource typing concepts,
- update utility emergency response plans,
- explore the exchange of resources across state lines through EMAC, and
- conduct additional training and exercises.

## LOCAL WARN INCIDENT RESPONSES

In addition to WARN responses to major regional disasters (e.g., earthquakes, tornadoes, hurricanes), WARNs across the country have been used to facilitate assistance during smaller or localized incidents that, in many cases, never necessitated a declared emergency or disaster declaration. The incidents described here have been organized into four categories and include examples of smaller or localized incidents to support them.

### Sharing Equipment

Equipment failures can occur during an emergency, and replacement equipment may not be stocked by the utility or there may be a long wait time to receive replacement parts from a manufacturer. Another WARN member may have equipment to share to shorten the response time. Following are examples of equipment sharing through WARNs.

CalWARN (California)

- The El Niño storm of 1997 caused significant flooding in California and subsequent landslides. Water transmission lines were disrupted in Monterey County; this affected water delivery to the county hospital, and plans were implemented to evacuate the hospital. CalWARN was activated to send portable pumps, aboveground flex pipes, and staff to respond and re-establish the transmission line. The water system, which relied mostly on pumps, was operational, and evacuation of the hospital was avoided.

KYWARN (Kentucky)

- A utility requested 14-in. ductile iron pipe after their pipe was washed out during a flood. Through KYWARN, the Kentucky Rural Water Association databases were queried, and KYWARN was able to find all utilities that had this type of pipe. The requesting utility received the needed pipe from a KYWARN member within a few hours, and service was restored in two days.
- A KYWARN member provided generators to Pikeville, Kentucky, after a flash flood. Among all WARNs, generators are commonly traded equipment.

MEWARN (Maine)

- On Sunday, December 23, 2012, the soda ash pump at the Alfred Water District broke down. The utility did not have any spare parts, so MEWARN was activated. Within 15 minutes, the Sanford Water District notified the Alfred Water District that they had an extra pump. The work was completed by Christmas Eve.

NEWARN (Nebraska)

- As a result of flooding on the Elkhorn River in June 2010, a 10-in. water main broke under the river, resulting in loss of the city of Stanton's primary water source. Within 24 hours, the city of Norfolk had assisted Stanton in installing fire hydrants on both sides of the river. The hydrants were connected by several hundred feet of large-diameter fire hose, which reestablished the line across the Elkhorn River bridge. This temporary connection was used for several months before a new water line was laid across the river.

- In June 2011, due to flooding on the Missouri River, Dakota City sent out a request on the NEWARN website for 4-in. or 6-in. trash pumps. The city of Norfolk responded with a 4-in. trash pump to assist Dakota City.
- In June 2011, a lightning strike damaged a control panel at one of the wells in the city of Humphrey. The damage to the control panel caused excessive pressure in the mains and contributed to six water main breaks in 10 days. Through NEWARN, the city of Norfolk provided Humphrey with a pressure relief valve for a nearby hydrant, which helped bleed off the excess pressures.

## Sharing Supplies

Disruptions that occur to supply chains can impact delivery schedules for water treatment chemicals and spare parts. In addition, local suppliers may not have the needed supplies due to other circumstances. Following are examples of supply sharing through WARNs.

SoonerWARN (Oklahoma)

- A utility needed a specialized replacement valve for a critical piece of equipment and found that their supplier would not be able to get them one for at least three weeks. The utility put out a request through SoonerWARN and was able to get the replacement valve from another member within four days.
- The only water well for a small municipality in northeastern Oklahoma was offline for almost a week in late 2008; its citizens were using bottled water. However, in the early afternoon of Christmas Day, the supply was dangerously low, and all area stores were closed due to the holiday. Through the Oklahoma Department of Environmental Quality hotline, the SoonerWARN administrator was contacted. After speaking with the town manager, a notice was posted on the SoonerWARN website, and e-mail messages were sent to all SoonerWARN members. Two SoonerWARN members responded with bottled water and, later in the day, hand-delivered the cases to the town and its citizens.

## Sharing Personnel

Extra personnel may be needed to respond to an incident or to fill in during some other type of utility personnel disruption. Following are examples of personnel sharing through WARNs.

CalWARN (California)

- In April 2010, a 7.2-magnitude quake occurred in southern California. CalWARN was activated to place representatives at the county and state levels to help coordinate information on the quake's impact on the water and wastewater systems. Personnel were deployed from member agencies to assist with damage and safety inspections, aboveground tank engineering assessments, treatment operations, and distribution support. Additional personnel were used for diving into the tanks and providing project management of the installation of temporary treatment systems. In addition, staff helped with the federal EMAC documentation needs.

MEWARN (Maine)

- In September 2012, MEWARN and the Maine Rural Water Association assisted the Canton Water District after they had a water main break under a brook that had emptied their reservoir.

SDWARN (South Dakota)

- Through SDWARN, personnel from the city of Ft. Pierre responded to Perkins County Rural Water after they had a leak and ran out of water in the city of Bison. The governor and the secretary of the South Dakota Department of Environment and Natural Resources sent letters and notes recognizing SDWARN for their response to South Dakotans in need.

### Growing Incidents

In some cases, incidents start out small and localized but quickly grow and require a multitude of resources. The case study presented here describes such a situation.

### Case Study

#### *Salmonella* Outbreak in Alamosa, Colorado, Water Supply

An outbreak of waterborne disease associated with *Salmonella* in drinking water struck Alamosa, Colorado, during March and April 2008. The city of Alamosa's public water system, which supplies drinking water to the community, became contaminated with *Salmonella* bacteria. Alamosa's population is about 8,900. The outbreak resulted in 442 reported illnesses, 122 of which were laboratory confirmed, and 1 death. Epidemiological estimates suggest that as many as 1,300 people may have become ill. Alamosa's drinking water comes from deep artesian wells in an aquifer considered to be a protected groundwater source. Before the outbreak, the city's drinking water was not chlorinated for disinfection. A statewide response to the outbreak, which lasted about one month and involved numerous responders, was coordinated using the National Incident Management System and Incident Command System. During the outbreak, residents were advised to drink bottled water. The entire water system was then flushed and disinfected with chlorine to kill the *Salmonella* bacteria. A boil-water order followed the bottled-water order and remained in place until tests confirmed that the city's water was safe to drink. Because of the quick operational response and the changes made to the physical infrastructure of the Alamosa public water system, the water in Alamosa has been safe to drink since April 2008. The city has since installed advanced treatment processes and improved system operations.

*Source:* Falco and Williams (2009)

## CONCLUSIONS

Recognizing key partners and developing relationships before an emergency are invaluable to effective response and recovery. Building familiarity and understanding of operational needs before an incident occurs facilitates quicker response times by all partners. MAAs, ideally established before an incident occurs, have proven to be important factors in the water sector's ability to maintain and recover services under a variety of scenarios and have benefited utilities of all sizes and types. The water sector has always leveraged this type of communal response, and initiatives such as WARN make the process stronger by establishing clear response protocols for the sharing of resources.

## REFERENCES AND SUPPLEMENTAL READINGS

AWWA. 2008. Water & Wastewater Mutual Aid & Assistance Resource Typing Manual. www.awwa.org/Portals/0/files/resources/water%20knowledge/rc%20emergency%20prep/rc%20warn%20resources%20pdf/typingmanual.pdf.

AWWA. 2004. Emergency Power Source Planning for Water and Wastewater Utilities. Denver, CO: AWWA.

Dappert, A.F. 1942. New York State Mutual Aid Plan for Water Service in Case of Emergencies. *Journal AWWA,* 34:2:189–195.

Devendorf, E. 1947. Mutual Aid Program for Peacetime [N.Y.]. *Journal AWWA,* 39:6:520–526.

Devendorf, E. 1943. Mutual Aid in New York in 1942. *Journal AWWA,* 35:2:201–206.

Falco, R. and Williams, S.I. 2009. Waterborne *Salmonella* Outbreak in Alamosa, Colorado March and April 2008—Outbreak Identification, Response, and Investigation. Denver, CO: Colorado Department of Public Health and Environment.

Geupel, L.A. 1943. The Mutual Aid Plan in Indiana. *Journal AWWA,* 35:2:227–231.

Geupel, L.A. 1941. Utility Co-ordination in Emergencies. *Journal AWWA,* 33:6:1079–1084.

Morley, K.M. 2015. WARN: A Partnership for Water Sector Resilience. Presentation at EMAC: Building National Mutual Aid Capacity Workshop, Minneapolis, Minn.

Morley, K. and Riordan, R. 2006. *Utilities Helping Utilities: An Action Plan for Mutual Aid and Assistance Networks for Water and Wastewater Utilities.* www.awwa.org/Portals/0/files/resources/water%20knowledge/rc%20emergency%20prep/rc%20warn%20resources%20pdf/WARNactionplan.pdf.

Pennsylvania Section of the American Water Works Association. 2011. PaWARN Fielded Many Requests for Assistance During Floods, Power Outages. *The Water News Source,* 47:3:8–9.

Potter, H.H.; Devendorf, E.; Murdoch, J.H., Jr. 1942. The Mutual Aid Plan at Work. *Journal AWWA,* 34:8:1173–1188.

USEPA (US Environmental Protection Agency). 2011. Planning for an Emergency Drinking Water Supply. www.epa.gov/sites/production/files/2015-03/documents/planning_for_an_emergency_drinking_water_supply.pdf.

USEPA. 2009. WARN Operational Plan. www.awwa.org/Portals/0/files/resources/water%20knowledge/rc%20emergency%20prep/rc%20warn%20resources%20pdf/WARNoperationalplan.doc.

USEPA. https://www.epa.gov/environmental-topics/water-topics

Whitler, J. 2009. Lessons Learned From WARN Tabletop Exercises. *Journal AWWA,* 103:12: 26–27.

Whitler, J. 2007. Emergency Preparedness for Drinking Water and Wastewater Systems. *Journal AWWA,* 99:3:36–38.

This page intentionally blank.

Chapter 5

# Internal and External Communications

This chapter discusses the various forms and methods of communication that are critical to utility operations and emergency response. Effective crisis communications require that a crisis communications plan (CCP) be in place before it is needed. The CCP should be a part of or in support of the utility's emergency response plan (ERP). While communication with customers, employees, and other stakeholders is always important to utilities, it is especially critical and demanding during an emergency. Every water and wastewater utility, large or small, must have an effective CCP.

In any emergency, a utility should be prepared in advance to communicate with internal and external stakeholders using a variety of methods. Customers expect service restoration as soon as possible after any water outage, from a main break to a plant shutdown. While this is not always possible, and certainly not within the high expectations of customers who are already under extreme stress, the utility should be prepared to answer questions about system recovery and service restoration with a sense of urgency, confidence, and understanding of the public's expectations.

## GENERAL CRISIS COMMUNICATIONS PLANNING GUIDANCE

One of the many important factors of resiliency is the ability to manage communications properly and appropriately during a crisis. Successful management of crisis communications involves having an agency-wide plan, identifying roles with specific individual responsibilities, and updating and practicing such plans with partners and stakeholders.

The three guidance documents that can assist a utility in the development and implementation of a CCP and/or in understanding the concepts involved include the following:

- This workbook, prepared by the California Department of Public Health, is for use by local community water systems in California (http://cdphready.org/crisis-and-emergency-risk-communications-toolkit-2/)
- Crisis and Emergency Risk Communication Resources. These are resources provided by the Centers for Disease Control and Prevention (CDC) (https://emergency.cdc.gov/cerc/index.asp).
- Drinking Water Advisory Communication Toolbox. This toolbox was created as a result of a joint effort of the CDC, AWWA, and US Environmental Protection Agency (USEPA).

Web addresses for these documents are listed in the References and Supplemental Readings section.

## Preparing Communications for a Crisis

Following is a review of what should be included in a CCP when preparing for a crisis and responding to a crisis.

Utilities should take the following steps when preparing for a crisis:

1. The person in charge of public communications, referred to as the *public information officer* (PIO), should be in control of the message. It does not have to be the "spokesperson"; ideally, they are one and the same, but not always. For example, the superintendent or chief executive officer may be in charge of the message and that individual may assign someone else to be the spokesperson.
2. Develop a CCP for use during a crisis.
3. Compile and maintain a list of contact telephone numbers (e-mail might not be available in a crisis) for each CCP contact. The contact information should include business, cell, and home phone numbers.
4. Identify utility personnel who can effectively speak in front of the camera. This assignment should be given to a person who can communicate clearly and effectively with the media. (See Understanding the Traditional Media section for more information on how to do this and who to consider.)
5. Notify all utility personnel that only designated personnel should speak with the news media, unless otherwise instructed.
6. Develop media (e.g., television, newspaper, and radio), community, and government contact information (e.g., business and cell phone numbers, fax numbers, e-mail addresses, and names of key contacts) and keep it current.
7. Create a primary location for the "media center" as well as a backup location. The media center should have the following:
   - appropriate supplies and equipment to handle electronic communication, faxing, phoning, and overnight mailing;
   - TV, radio, and VCR/DVD capability for monitoring media coverage and recording it;
   - provisions for media briefings, including a podium with an appropriate backdrop and a public-address system, if needed;
   - a work area with phones for the press; and
   - appropriate support staff and assignments to help the PIO produce needed media information.

8. Involve the public to determine which groups might require specific direct communication and if the CCP can be improved to best serve customers' health and safety needs in times of crisis.
9. Prepare response sheets for customer service telephone operators.
10. Maintain a utility web page for ongoing stakeholder communications.

## Communicating During a Crisis

Utilities should take the following steps during a crisis:

1. Establish a clearinghouse for reporting response and recovery progress, delays, and new issues for the PIO.
2. Document everything known about the crisis in writing for review. Update the report regularly.
3. Determine what key points should be made available to the media.
4. Immediately after review (within 1 hour of the event), develop and distribute an initial media statement. Immediately update the utility's website with this and future press releases. Provide the media with the PIO's name and phone number to streamline inquiries. Implement the regulatory-required drinking water advisory process.
5. Activate the media center. Communicate activation with persons charged with security and implement appropriate access rights. The utility should have a system for credentialing approved media.
6. Contact external public relations firm, if appropriate.
7. Communicate individually with stakeholders specific to the utility, including:
    - utility employees,
    - first responders,
    - governmental entities and elected officials (notify first, if feasible),
    - customers and the public at large,
    - the media, and
    - others including the medical community, special interest groups, and others as identified in the public involvement effort.
8. Maintain files of all hard copy, electronic, media, and video communications during the crisis event.
9. Provide the media center support staff with lists of pertinent/appropriate contacts and telephone numbers and contacts. These should be compiled in advance and accessible from the CCP.
10. Provide frequent updates to employees regarding the crisis event and the utility's responses and press releases.
11. Direct all employees to refer media requests to the designated spokesperson or PIO.
12. Provide media with background information, including information about how the utility protects public health and safety. This could be part of a press kit.
13. Request support from partners such as AWWA, the Water/Wastewater Agency Response Network, and other appropriate federal, state, and local organizations.

14. If activated by city, county, state, public health, or other agencies, communicate with the joint information center (JIC) and coordinate on the delivery of public messages.

The PIO should be in control of the message that is being conveyed by doing the following:

- Anticipate likely questions. Understand the public's perceptions and fears. Address fears by offering facts, not conjecture. Do not answer "what if" questions. Do not use emotional statements or industry jargon.
- If it is necessary to communicate with the public before all facts are known, be humble, state priorities (e.g., getting the situation under control), and assure the audience that the utility staff is doing everything they can.
- Leverage message mapping resources that provide guidance in developing the framework of key statements before the events (USEPA 2008).
- Recognize that there is no such thing as "off the record."
- Avoid conjecture and assigning blame.
- Keep communications brief and concise to minimize confusion and dilution of the main message.

## INTERNAL COMMUNICATIONS WITH EMPLOYEES

During a crisis, it is important to communicate with the staff and inform them of the situation and what they need to know to respond in an appropriate manner. An important part of crisis communications is communicating with utility employees. Planning efforts should consider how the utility will communicate with employees during and after an emergency, how emergency preparedness expectations will be incorporated into new employee orientations, and how the utility will develop and provide training and exercises. These internal communications should do the following:

> ***ANSI/AWWA G440, Sec. 4.6.1:*** *Regular and ongoing communications with employees. The utility shall establish, document, and maintain strategies for effective communications with employees about preparedness issues.*

- convey the employees' expectations during emergency and disaster incidents;
- provide two-way, that is, inbound and outbound, communications systems for employees;
- designate which employees should report to work based on the incident and where they should report;
- provide default guidance to employees in the event of communications failure; and
- help employees and their families prepare for emergencies and disasters.

## COMMUNICATING WITH EMERGENCY RESPONSE AGENCIES AND LOCAL GOVERNMENT AGENCIES

Communicating with first responders and local government agencies before any emergency occurs is an important part of the emergency preparedness program. This communication takes place before any incident occurs. The best place to get to know interagency counterparts is in a nonemergency environment. Ways to help facilitate this interagency familiarity include the following:

- hosting interagency tabletop exercises or other types of exercises and inviting the other agencies,
- getting the utility invited to participate in or observe exercises conducted by the other agencies,
- inviting representatives of the other agencies to meetings,
- offering other agencies the utility's facilities for training and meetings, and
- working with other agencies on the concepts, actions, and resource needs when creating a JIC, where collaborative media messages can be drafted and shared with the media.

In an emergency, utilities should communicate first with other, more traditional first responders including fire, rescue, law enforcement, emergency management, and health agencies, as well as the local government agencies of which they are a part. Communications with these agencies generally takes place in one of the following three ways during an incident:

- *9-1-1:* In the event of a serious incident that begins at the utility, the initial communication is a call to 9-1-1. This form of communication comprises a small portion of the total communications effort.
- *On-scene:* A major form of communication with other first-responder organizations is face-to-face, on-scene communications. It is important for utility representatives to understand their roles and responsibilities while on the scene. If the incident is a potential crime scene, law enforcement will be the lead agency, even if the event occurred at a utility facility. If the incident involves a fire, hazardous materials incident, structural collapse, or other hazardous condition, the fire department will be the lead agency.
- *Emergency operations center (EOC):* The emergency management agency is a critical link in the communications with county, state, and federal response agencies. In a major incident, the primary and formal line of communication between agencies will be through the incident command system (ICS) at the EOC. It is critical that utility representatives be trained in and familiar with ICS procedures and terminology to effectively participate in this environment. Most EOCs are hosted by the appropriate emergency management agency, generally at the local, county or parish, and state levels.

## COMMUNICATING WITH CUSTOMERS AND THE PUBLIC AT LARGE

Effective public information and communications during an emergency are crucial. The CCP must be specific to the utility, and the utility should take the following steps when developing their plan:

- Incorporate communication and preparedness into the organizational culture. Utility personnel need to know who they are to communicate with and the methods to be used under normal circumstances and during an emergency.
- Designate the PIO and any additional spokesperson, each of whom performs a very important role in managing and communicating information to the public and concerned parties. The PIO, through the local media center or JIC (if activated by other agencies), communicates with the media.

> ***ANSI/AWWA G440, Sec. 4.6.3:** Regular and ongoing communications with customers. The utility shall establish strategies for effective communications with customers prior to any emergency. Communications strategies should consider the most effective way to reach customers with information, both in terms of delivery and source, and ways to get information from customers on unusual events or suspicious activities. The utility's strategies should consider key messages, which person is trained and designated to deliver the messages, and the need to be first, be right, and be credible in public communications, especially during an incident.*

- Evaluate scenarios that may adversely affect communication systems, and develop processes to prevent or mitigate effects or implement backup and alternate systems.
- Prepare to communicate early and often following an emergency or disaster. Messages should be provided at least twice per 24-hour period and whenever there is a notable change in information.
- Develop the CCP before a disruption or contamination at a utility. The CCP should include modes and timing of communications with utility customers. Customers fall into different categories depending on how they use utility services and the extent to which their operations are dependent on the services delivered. Examples of users that are critically dependent on utility service delivery include hospitals, schools, fire departments, and food and beverage producers. Communication with the general public about health and safety risks is also critical.
- Provide mass media, particularly television and radio, with vital information about what actions to take and how to protect themselves.
- Make special provisions to communicate with special-needs populations including:
  - customers critically dependent on water, such as dialysis patients;
  - the frail and elderly;
  - pregnant women, infants, and young children;
  - non-English-speaking populations;
  - customers without any form of electronic communications; and
  - homeless populations.
- Consider involving the concerned public in a review of the CCP to build a constituency of credible sources with whom to partner in case of crisis. Potential candidates for credible sources include local universities, parent–teacher organizations, medical professionals, environmental groups, and nongovernmental organizations.

## Designating a PIO and Additional Spokespersons

Disasters and emergencies are newsworthy. In times of crisis, it is essential to designate the correct person to provide the public with accurate information. All information should be presented with a tone of urgency, compassion, and understanding. The person should convey an accurate and honest characterization of the situation, demonstrate the utility's competence to remedy the situation, and honestly communicate the facts and honest expectations with confidence and assurance.

The PIO, who is the agency representative responsible for coordinating media communications, bears the critical responsibility of expediting effective and accurate communications and disseminating this information appropriately. The PIO should possess knowledge about media communications, disaster management, the utility's operations, and the capabilities of any other designated spokespersons within the utility. The PIO should have the ability to work under pressure and think quickly on his feet.

Every staff member should have a full understanding of the PIO's role as well as the internal procedures for referring members of the media to the PIO. Operating within the ICS framework, the PIO answers to the incident commander in the field or to management in the EOC. All requests by the media are made directly to the PIO or referred to the PIO.

The PIO's tasks include the following:

- developing and maintaining a media contact list,
- selecting appropriate team members for interviews,
- scheduling interviews,
- preparing selected team members for interactions with the media,
- escorting media representatives and VIPs throughout the operational area,
- issuing press releases,
- gathering and verifying information, and
- briefing a spokesperson, if the PIO is not in front of the media.

There are times when the PIO is not the person in front of the media. For example, the general manager or utility's chief executive officer may address the media, with support from the PIO. Other utility staff with expertise on the subject at hand, e.g., water quality or an effluent release, may assist the person in front of the media. The key is that these individuals are supported by the PIO and prepared to speak in front of the media to explain the situation with confidence. An effective spokesperson should:

- Be prepared in advance and trained to answer difficult questions;
- Understand the need to focus on what should, and should not, be communicated;
- Maintain a positive, forward-looking attitude, even when criticized, and never appear to be on the defensive;
- Speak with one voice no matter what may be occurring internally;
- Understand that they are speaking for the utility and not themselves;
- Know how to bring other experts into the communications mix when their expertise is warranted;
- Understand the importance of frank and honest dialogue; and
- Understand the public's anger and frustration with the crisis, communicate empathy, respect the media's demands and comply with media deadlines.

## Public Information Training

It is critical that the PIO and all other spokespersons be trained on writing messages, presenting in front of media, and presenting the key message required to help the customer understand the issue and act based on the information. Training is available from many resources including AWWA, federal emergency management agencies including the Emergency Management Institute, the USEPA, and professional organization conferences and workshops. For more information, see the References and Supplemental Readings section.

## Preparing Messages

When a crisis or disaster occurs, the utility, as a provider of essential services, should be prepared to provide regular and clear information regarding the utility's response and recovery efforts. Under these circumstances, utility staff should recognize that because of high stress levels, the public does not always comprehend or retain much of the information that is provided. Consistency and repetition of open and empathetic information that includes acknowledgment of the situation's uncertainty as recovery efforts continue is the best messaging method for a utility.

One technique for preparing draft messages in advance is to review vulnerabilities and the impact that catastrophic events may have on the utility's ability to service customers immediately following an outage that could last for one day or for weeks. The utility should consider the operational impact from an extended power outage, loss of communications, or a contamination event and the types of questions that these events would generate from customers, public officials, and the media.

The PIO and others in the utility should develop messages that convey important facts related to the event before answering questions and should do the following:

- State the most important messages up front and repeat them as often as possible.
- Repeat and emphasize critical information because information overload and the stress of the situation can make it difficult for the public to comprehend the situation.
- Develop message templates for the types of hazards the utility has identified in their risk assessment.
- Learn and understand the technique known as "message mapping" to anticipate questions and develop appropriate answers. Information on message mapping can be found in USEPA guidance documents (USEPA 2012, 2007a, 2007b).

Messages and questions tend to fall into specific categories. While it is impossible to know in advance what questions will be asked, preparing responses in specific categories of interest can be very helpful. For example, questions for a utility in a disaster situation would include:

- What happened to cause the emergency or system failure? Provide specifics as to how and why.
- How many customers are impacted? Either provide specific numbers or discuss assessment procedures.
- When will full or limited service be restored? Discuss assessment and recovery operations. If major system damage has occurred, be frank and open about the extended periods of downtime.
- Is the water safe to drink, to use for bathing, and for other uses? Discuss restoration and testing, boil-water notices, and other steps that can be taken.
- Where can the public get water?

## Delivering the Message

How the message is delivered is as important as what the message is. The spokesperson or PIO who conveys the message should:

- Convey compassion and empathy for all stakeholders including employees and their families.
- Realize that in a disaster, no one is immune from the impact. Everyone is pulling together to recover as quickly as possible.
- Not be defensive or convey body language that is anything less than open and transparent.
- Never get angry, even when challenged. It is a reporter's job to get information. Their efforts are not a personal attack.

- Not be nervous or allow nervous laughter or fidgeting to occur as it undermines credibility and is inappropriate in a disaster situation.
- Stay in control and show calm under the stress of the situation. This provides confidence that the situation is under control and maintains the appropriate tone of the situation.

## UNDERSTANDING THE TRADITIONAL MEDIA

Traditional forms of media, which include print, radio, and television, are rapidly evolving as social media changes the way information is presented to the public. Media outlets rely on social media as well as traditional media to disseminate information. In the past, each outlet had reporters, most with journalism training, predictable deadlines, and were operational during normal business hours. Today, media includes all forms of social media that operate with instantaneous message delivery. Today, all media outlets have websites and use social media to gather and disseminate information. Sometimes this information is crafted into a full story. Further discussion of social media is presented in the following section.

Although the task of responding to media outlets in a crisis is daunting, it is as essential to the restoration and recovery process as any of the utility's other responsibilities to its customers. Knowing how to respond to the media is essential.

As the provider of an essential service, utility leaders cannot refuse to speak to the media. Sometimes the designated PIO works with the media; at other times, it is essential that the utility leadership appear before the media to give statements and answer questions. Therefore, it is critical to have a well-rehearsed CCP as part of the utility's ERP and it is important to understand the media.

Following is a brief overview of the traditional print, radio, and television media. It is important to keep in mind how these three overlap and that, at any given time, all three will be the utility's audience and what is said will be communicated to the customers. Most reporters and media representatives are not utility experts, so it may be necessary to explain how the utility operates as well as recovery actions. Questions that may seem simplistic to the utility staff can be expected. It is important to help the reporter put the story in context. Spokespersons should also expect questions that may seem to challenge the utility's level of preparedness for a disaster. At times, this may appear to be an attempt to add drama to the story. Because of the highly competitive nature of news-gathering and dissemination, utility leadership should anticipate this happening.

### Print

Print media is becoming less and less relevant in some cities. The one important characteristic of print media is the ability to provide detailed coverage of events every day and to provide greater detail to complex stories. Because of competition with other forms of media, including social media, many print outlets are now available online and update their stories as needed.

### Radio

Immediacy is radio's advantage. Often after a disaster, radio is the first to present the story. However, with the advances in television technology, newscasters are often on the scene quickly and can present information in real time. Radio usually has multiple deadlines throughout the day, so interviews may be live or recorded and then repeated with every news cycle on the hour or half-hour. With radio, regular programming can be interrupted with updates.

## Television

Television requires visuals to be effective. In the past, this meant having to video tape events. However, during the initial hours and first days of a disaster, TV can go live and create a story around the available visuals. Good journalists always tell the story based on the facts they have gathered. Complex stories with ever-changing details, even with the events unfolding live on screen, are difficult for journalists to convey in a way that the general public will be able to follow.

Cable and local news outlets operate 24/7 and, by their very nature, look for new angles and directions that a story may take during the day. Many operate their news cycle in hour-long blocks or portions of an hour so that stories can be repeated multiple times within an hour and then updated as needed each hour, or repeated until news changes. Such a change might be presented via a new press conference or a status update of recovery operations.

## Preparing to Meet With the Media

In preparing to meet with the media, the spokesperson and PIO should:

- Know the open records law in their state;
- Know what should be said and can be said and what information should be kept confidential;
- Not attempt to keep information confidential unless there is a valid reason, such as security issues;
- Be aware of the Health Insurance Portability and Accountability Act regulations when discussing employees and the public;
- Know the media, anticipate their questions, and prepare and rehearse the responses;
- Not talk to reporters until they are ready;
- Be prepared to answer questions on the media's schedule;
- Always find out the reporter's name and their news organization; this can be done using utility-issued media ID badges, if possible;
- Practice the message;
- Realize that notes are acceptable and there is no need to hide them; an outline instead of a well-crafted statement can be used;
- Remember that they are communicating information, as well as the utility's sincere efforts to restore service;
- Not be led into responding to leading questions that make false assumptions. Learn the technique of rephrasing the question and answering it as it should be answered from the utility's perspective;
- Be consistent with messages that have already been provided;
- Openly correct any misinformation and describe situational changes;
- Remember that the utility did not create the disaster and that the public is not holding the utility responsible for the event;
- Do their best to accommodate reporters' needs while maintaining an environment in which the spokesperson and PIO are comfortable;

- Be fairly treated, with respect and courtesy;
- If interrupted, politely state that they are going to finish their remarks;
- Give all reporters the chance and time to ask questions when speaking with several reporters at the same time;
- Record all interviews to avoid being quoted inaccurately and let reporters know that the meeting is being recorded;
- Focus the interview. Do not let a reporter ask about a broad topic and expect an in-depth answer;
- Not be intimidated or pressured into an answer. It is okay to not know every answer, particularly during disaster recovery. Let the reporter know they will get back with an answer and then do so;
- Not be afraid to stop if they realize they said something wrong. Keep the record accurate;
- Be prepared to educate the reporter about what the utility does and why;
- Avoid jargon and too many technical details. At the same time, explain how the recovery steps will restore some semblance of normal operations; and
- Avoid saying "no comment."

## USING SOCIAL MEDIA

Social media has developed into the fastest, most far-reaching, and rapidly evolving tool to use when providing utility customers with the information they need during disaster recovery. Social media often "broadcasts" messages about an emergency before the utility even knows one exists. Competent social media management is a critical part of a utility's communications strategy and is as important as a press conference. In fact, during a disaster, the public is more likely to follow and trust messages on social media than other sources.

Often, video clips are provided via social media without a story and without context. Additionally, the public has embraced the use of social media to show personal camera shots and videos obtained via cell phones. This can occur more often during a disaster, and the visuals may tell a story that is not accurate or representative of the recovery situation. Also, because the news media routinely use social media as a source for tips and story lines, it is important that utilities and all responders know how to track and use the information being communicated using social media.

Most utilities are familiar with the use of public safety communication systems such as Reverse 911 to communicate with customers. In fact, many utilities have incorporated these systems as one of the most immediate forms of communications.

The reach of cell phones and tablet devices is becoming nearly universal. Nearly half of all citizens have immediate access to at least one of these communication devices. Most devices can be used to take reasonably high-quality still pictures and movies and to send these images along with messages via e-mail, text, or through social media sites such as Twitter and Facebook. The devices are also able to receive messages and images sent by others.

Based on past experiences during disaster events, text or short message service (SMS) messages from wireless communication devices is possible even if the cellular signal is not strong enough to make a voice call. This is because a different technology is used to send text messages through the SMS wireless gateway. First responders increasingly rely on their ability to send SMS messages to mobilize disaster relief personnel and coordinate disaster relief operations.

Utilities should consider using social media as a means of communications during response when:

- providing emergency communications and issuing warnings,
- receiving requests for assistance,
- alerting utility staff to certain situations by monitoring the flow of information from different sources during an incident,
- providing immediate and direct communications to customers outside of the normal media channels or scheduled events,
- monitoring and receiving information on situational awareness, and
- using uploaded images to perform onsite damage assessments.

Social media can also be used to track what is being heard "on the street" and to determine if the utility's messages are being understood by the public. Used correctly, social media can help with rumor control and provide a trusted source of information. Additionally, utility staff can easily provide field information and damage assessments on a real-time basis, thus shortening the recovery response.

The decision to use and implement a CCP that relies on social media must be well researched and implemented with a full understanding of the complexities involved. The use of social media requires a commitment by staff and equipment that is dedicated to developing, transmitting, and gathering information in an organized 24/7 manner for the critical 72- to 96-hour period after a disaster occurs.

An increasing number of resources are being developed to provide guidance on the use of social media during a crisis. Ideally, use of social media is coordinated with the utility PIO so that messages are consistent, and the utility continues to speak with one voice. Key planning steps include:

- Developing the use of social media before a disaster occurs so the public will know where to turn for information. Begin by using social media to relay routine information about utility operations such as water conservation messages or how to resolve billing inquires.
- Committing staff and resources on a scalable basis to plan and execute the social media program;
- Recognizing that use of social media resources allows the utility to manage the messaging and story and to correct inaccurate information in real time;
- As part of the utility's ERP, creating social media use policies and procedures that include:
    - Setting up a monitoring schedule;
    - Understanding the use of Twitter hash tags for directing messages and monitoring these communications regularly;
    - Training all staff in the utility's procedures on use of social media for sending and receiving messages. This includes designating staff with the authorization for messaging to the public, field operations for damage assessment, and similar recipients.
    - Developing procedures to ensure that the PIO, spokesperson, and leadership team speak with one voice using all types of media;
    - Creating web pages and social media sites that are "disaster ready" and that are normally "dark," yet can be launched within minutes of a disaster;
    - Developing a messaging tree that provides clear guidance on how and when to send and respond to messages. A well-thought-out response is better than a rapid response. Everything said online has a long "shelf life."

- Ensuring that transparency is part of all messaging, including the use of social media, and that all messages communicate the tone of compassion and honesty necessary in a crisis.

## COMMUNICATING WITH REGULATORY AGENCIES

> ***ANSI/AWWA G440, Sec. 4.6.4:** Regular and ongoing communications with regulatory agencies. The utility shall establish strategies for effective communications with relevant regulatory agencies. Communication strategies should consider timely two-way communications in the event of an actual incident or threat.*

Local, city, county, state, and federal regulatory agencies are an important source for resources necessary during a crisis. Regulatory agencies can provide direct assistance and technical advice on water issues, perform assessments and inspections, supply emergency water quality sampling kits, and provide laboratory support and additional technical support. For these reasons, it is important to include regulatory agencies among the communication partners in the CCP. Figures 5-1 and 5-2 depict how such communication can facilitate resource requests.

A utility should include regulatory agency contact information in the CCP and keep it updated. Also, the utility should establish strategies for effective communications with relevant regulatory agencies. Communication strategies should consider timely two-way communications in the event of an actual incident or threat.

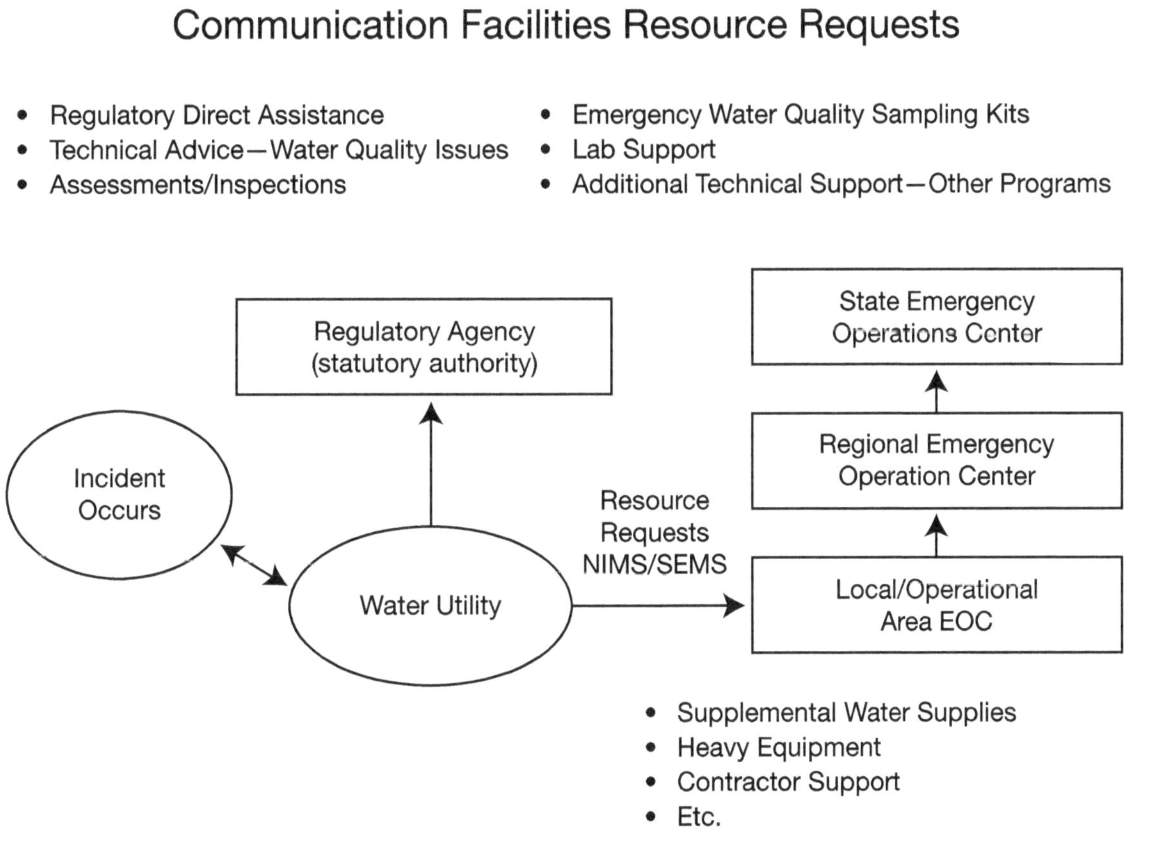


EOC: emergency operations center, NIMS: National Incident Management System, SEMS: Standardized Emergency Management System

**Figure 5-1 Communication facilitates resource requests**

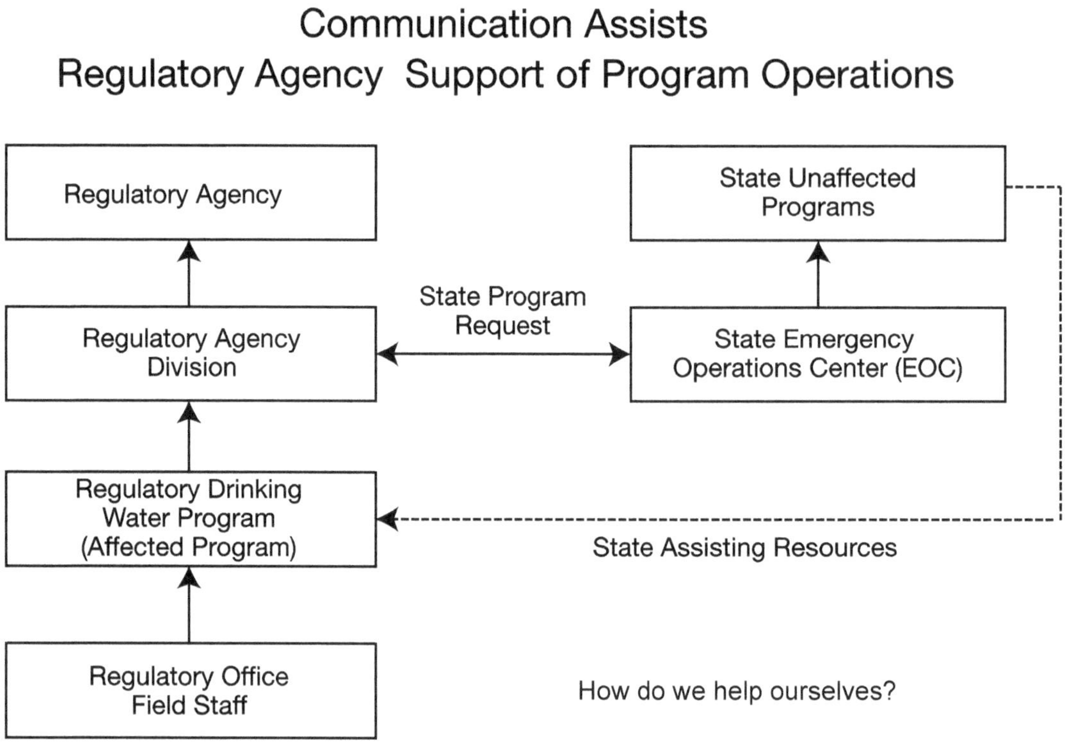


Figure 5-2 Communication improves regulatory agency support of program operations

## MAINTAINING CONTACT INFORMATION

A communication chart should list all utility staff and alternates who will direct emergency response actions, as well as all external contacts that may be needed. The chart should include job titles, telephone numbers, and major responsibilities. Others to include on the communication chart include additional personnel, other utilities, regulatory agencies, priority customers, and media contacts for general public notification. Alternative communication methods, such as radios and cell phones, should be included if normal telephone communication is disrupted.

## CONCLUSIONS

Effective crisis communications require that a CCP be in place before it is needed. It is important to periodically update and exercise the CCP. This should also be done after lessons learned from exercises and real events. See Chapter 6 for more information about exercises.

It is critical that utilities communicate in a way that fosters trust and confidence. The utility should be the first to release information about how the crisis has impacted service delivery. Delayed release of information can result in loss of life, accusation of a cover-up, and ultimate loss of public trust. In a crisis, the guiding principle for utilities is, "be the first, and be right."

Utilities that have CCPs and that designate and equip their PIO and spokesperson will be much better prepared for and successful in communications with their stakeholders during and after an emergency.

## REFERENCES AND SUPPLEMENTAL READINGS

AWWA. *Public Communications Toolkit.* www.awwa.org/resources-tools/public-affairs/communications-tools/public-communications-toolkit.aspx#2213179-digital--electronic-communications.

California Department of Public Health. 2011. *Guidance Documents for Development of a Crisis Communications Plan.* www.cdph.ca.gov/certlic/drinkingwater/Documents/Security/CERCtoolkit.pdf.

Centers for Disease Control and Prevention. *Crisis & Emergency Risk Communication (CERC).* emergency.cdc.gov/cerc/books.asp.

Federal Emergency Management Agency and Emergency Management Institute. training.fema.gov/emi.aspx.

USEPA (US Environmental Protection Agency). 2012. Need to Know: Anticipating the Public's Questions During a Water Emergency. EPA/600/R-12/020. cfpub.epa.gov/si/si_public_record_report.cfm?subject=Homeland%20Security%20Research&dirEntryId=240476.

USEPA. 2007a. Effective Risk and Crisis Communication During Water Security Emergencies: Summary Report of EPA Sponsored Message Mapping Workshops. EPA/600/R-07/027. www.epa.gov/sites/production/files/2015-08/documents/effective_risk_and_crisis_communication_during_water_security_emergencies.pdf.

USEPA. 2007b. Risk Communication in Action: The Tools of Message Mapping, EPA/625/R-06/012. http://nepis.epa.gov/Adobe/PDF/60000IOS.PDF.

This page intentionally blank.

# Chapter 6

# Training and Exercises

Once an emergency response plan (ERP) is in place, a utility should train and exercise staff, so they can practice preparing for and responding to incidents. This will build the skills needed to minimize losses and expedite recovery. Staff will gain confidence for responding to emergencies in a safe and efficient manner when they learn and practice their roles and responsibilities under the ERP. Other reasons to train and exercise include the following:

- testing and evaluating emergency plans, policies, and procedures;
- revealing gaps in resources;
- improving organizational and external coordination and communications;
- building relationships with internal and external partners;
- supporting an emergency preparedness culture;
- clarifying roles and responsibilities;
- improving individual performance;
- gaining program recognition and support of officials/customers; and
- satisfying regulatory requirements.

## TRAINING AND EXERCISE MATRIX

> ***ANSI/AWWA G440, Sec. 4.7.1: Training Matrix.***
> *The utility shall develop a training requirements matrix for employees. At a minimum, employees with roles in the preparedness plans must be trained on their organization, content, and use. Employees should be trained on NIMS and the ICS. Other staff may be trained on general emergency preparedness in accordance with maintaining an overall preparedness program.*

It is recommended that a utility develop a training and exercise matrix (Figure 6-1) to document and manage training and exercise goals. This matrix is a planning document that should be coordinated with the utility's risk assessment report and ERP, as well as related plans (e.g., business continuity plan, crisis communication plan).

In the training portion of the matrix, ensure that training activities are position specific.

| ABC Water Utility<br>Emergency Preparedness Training Matrix | | | | | |
|---|---|---|---|---|---|
| Position | Training Courses | | | | |
| | Course 1 | Course 2 | Course 3 | Course 4 | Course 5 |
| Job title 1 | Annual | Annual | X | Quarterly | Annual |
| Job title 2 | Annual | Quarterly | Annual | Monthly | x |
| Job title 3 | | x | x | Monthly | x |

**Figure 6-1 Example of an emergency preparedness training matrix**

Personnel should be trained based on the roles and tasks they perform when responding to an emergency per the ERP. When developing the matrix, consider training activities that support preparedness, response, and recovery. An example of preparedness training is what to do when a hazard is approaching, such as a hurricane or a period of extreme cold, to minimize losses. Examples of response training include training in incident management and public information. Examples of recovery training include how to return a system from emergency operations back to normal operations and how to perform a damage assessment or an incident investigation.

In the exercise portion of the matrix, ensure that each planned exercise is based on predefined objectives. These objectives may be developed by reviewing the priorities identified in the risk assessment report, response plan, internal meetings, training activities, incident investigation reports, and conversations with regulatory agencies and other government representatives about top risks to water utilities in the region. For example, a top threat might be cyberattacks. In this case, counteracting and responding to cyberattacks should be included in the utility's exercise program. The same is true for any top threats and hazards a utility faces. It is also helpful to consider recent emergency preparedness training activities when developing exercise objectives. Performing these activities soon after training reinforces the training and is very beneficial to a utility's emergency preparedness program.

The combination of training and exercises promotes a process of continual improvement and a preparedness culture. Another consideration for successful training and exercises is to perform these activities with other response agencies. A carefully developed multiyear training and exercise matrix will serve as a road map for greater preparedness.

The US Environmental Protection Agency (USEPA) developed a step-by-step guide for developing a multiyear training and exercise plan (USEPA 2011). According to this guide, the plan should consider including the following components:

- Purpose,
- Points of contacts,
- Program priorities and capabilities-based training,
- Methodology and improvement tracking, and
- Multiyear training and exercise plan schedule.

The following additional factors should be considered when developing a training plan:

- Planning should be conducted with the utility's response partners such as the local fire department(s), police department(s), public health department, local emergency management agency (EMA), state primacy agency, utilities in neighboring jurisdictions, and state Water/Wastewater Agency Response Network (WARN) members.

- To the extent possible, use examples and case studies that are pertinent to the utility. This will allow the trainees to better appreciate the concepts, particularly those that pertain to the incident command system (ICS) and the National Incident Management System (NIMS). For example, a case study on a water contamination event may be more effective than discussing the attack on the World Trade Center.
- It is important that utility management support the training and exercise plan. Coordination and execution of the plan should be assigned to one staff member.
- Ensure that utilities use their operating budgets to fund the training and exercise plans. Federal, state, and regional grants are sometimes available to conduct exercises. In addition, utilities should look for opportunities to join exercises conducted by neighboring water systems, WARN members, and local and regional EMAs.
- In addition to the planned exercises, view every real-world incident as a learning opportunity. Conduct a debriefing session, or "hot wash," after each incident and follow up on the improvements (action items) identified.

A template of a multiyear training and exercise plan is shown in Figure 6-2.

## INITIAL AND REFRESHER TRAINING

A successful emergency preparedness training program should cover initial training and refresher and ongoing training, as shown in Figure 6-3. Initial training is for new employees and newly promoted employees; refresher training is for employees who have been previously trained.

### Initial Training

Initial training has two components: orientation and specialized training, each with different objectives.

**Orientation.** Orientation training simply outlines the ERP to the staff. It describes the plan's components, emergency procedures, and the employees' roles and responsibilities. This training should be given to new employees and to all employees after the ERP has been revised significantly. Contract personnel with emergency preparedness responsibilities should attend the training. Senior management participation is essential to demonstrate the program's importance.

> ***ANSI/AWWA G440, Sec. 4.7.2:*** *Initial training shall be conducted for new employees with specific roles in preparedness during the initial orientation or prior to being given such a responsibility. Refresher training should be conducted at least annually, or more often as dictated by new threats to the organization, or following an incident where lessons learned should be immediately incorporated into overall preparedness.*

An orientation can include the following:

- Review of assigned responsibility and need to respond,
- Review need for training to do assigned task and meet requirement of ICS,
- Review of how to be prepared at home, and
- Review of how to be prepared at work.

At the end of the orientation training, the utility can collect information on the special skills each employee may have outside of work, such as being a search-and-rescue volunteer or having another specialized skill.

**Specialized training.** Specialized training addresses the skills, knowledge, and procedures necessary to execute the ERP. This training can cover a wide array of topics depending on the utility's needs. Common specialized training needs for most utilities

| Agency/ Jurisdiction | Medium Utility Multiyear Training and Exercise Schedule: Year One | | | | | | | | | | | |
|---|---|---|---|---|---|---|---|---|---|---|---|---|
| | Quarter 1 | | | Quarter 2 | | | Quarter 3 | | | Quarter 4 | | |
| | J | F | M | A | M | J | J | A | S | O | N | D |
| Medium utility | | Exercise: Coop workshop | | | Exercise: ERP workshop | Training: Telephone threats | Exercise: Emergency contact seminar | Exercise: Coop tabletop | | Exercise: All-hands ERP seminar | | |
| Neighboring utility name | | | | Exercise: Mutual aid workshop | | | | | | | | |
| County health department | | | | | | | Exercise: Pandemic influenza tabletop | | | | | |
| State primacy agency | | | | | | | | | | | | Exercise: Water security webinar |
| Local emergency management agency | | | Training: ICS-100.A | | | | | | Training: National Incident Management System ICS-700.A | | | |
| Local fire department | Training: Hazmat safety | | | | | | | | | | Exercise: Hazmat spill drill | |

Priorities addressed: utility training and exercise priorities (medium utilities can conduct some exercises and training as well as participate in those sponsored by their response partners)

Priority 1 **Safety** | Priority 2 **Communications** | Priority 3 **Continuity of Operations** → Example Utility Priorities

ERP: emergency response plan, ICS: incident command system
*Source:* USEPA (2011)

**Figure 6-2 Example of a multiyear training and exercise plan**

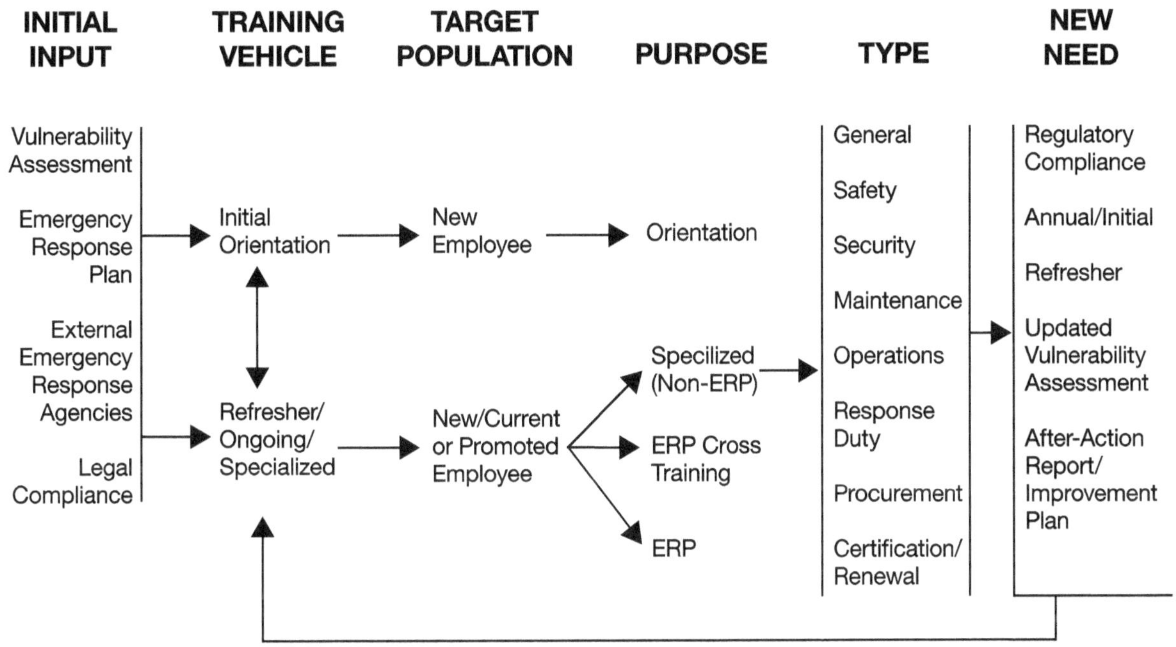


ERP: emergency response plan

**Figure 6-3 Emergency preparedness training plan model**

include incident management, operating an emergency operations center, evacuation and sheltering in place, communicating with the media, and managing incidents jointly with other first-responder agencies. Other specialized training needs include emergency shutdowns or emergency operations of critical systems, responding to a water system contamination event, regulatory compliance during specific emergencies, and changing the operation mode of critical systems (e.g., automatic to manual).

Concurrent training and/or emergency response management training can be provided. Concurrent training covers both routine and emergency response operations such as security, public information, and safety. Emergency management training teaches the necessary regulatory compliance and decision-making skills and knowledge needed to respond to any threat or emergency. The intended audience is employees who have direct emergency response duties and are vital to addressing an emergency. It is key to schedule initial training promptly for newly promoted employees with new emergency response duties and to implement a cross-training program to ensure that there is adequate staff to address absences due to adverse weather, illness, vacations, and position vacancies.

### Annual Refresher Training

Refresher training recalls and reinforces previously acquired knowledge that can be forgotten if not practiced or used regularly. Refresher training maintains staff's response effectiveness to current and new threats and hazards, new regulatory requirements, operational and procedural changes, and identified deficiencies from actual incidents, exercises, and drills in a dynamic environment. This training provides the opportunity to cross-train selected employees to ensure adequate response capability at all times.

## TRAINING SOURCES AND METHODS

When the training matrix identified in Figure 6-1 is completed, the next step is to schedule the training from internal and/or external sources. Internal training sources include current

**Table 6-1 Training methods and their general parameters**

| Training Method | Objective | Participation | Planning Effort | Location | Travel Expenses | Information Technology Support | Schedule |
|---|---|---|---|---|---|---|---|
| Online | S | Individual | Low | Anywhere | No | Yes | Flexible |
| PC-based (IH) | S | Individual | Low | Onsite | No | Yes | Flexible |
| Correspondence | S | Individual | Low | Anywhere | No | Maybe | Flexible |
| Meetings, one-on-one | S | Individual | Medium | Anywhere | Maybe | Maybe | Flexible |
| Meetings, video conference | S | Group | High | Anywhere | No | Yes | Set |
| Anywhere | S | Group | Low | Local | No | Maybe | Set |
| Seminars, offsite | D | Group | Low | Anywhere | Maybe | Maybe | Set |
| Seminars (IH) | D | Group | Medium | Anywhere | Maybe | Yes | Set |
| Workshops (IH) | D | Group | Medium | Anywhere | Maybe | Yes | Set |

D: discussion based, IH: in-house, S: skill or knowledge need for emergency response plan

staff and utility resources, from classrooms to course handouts and away-from-duty time. External sources can require additional funding and staff away-from-duty time.

The USEPA and Federal Emergency Management Agency (FEMA) websites are excellent sources for online training courses such as the independent study (IS) classes and program manuals and user guides. Today's reliance on the Internet for information dissemination makes it mandatory that all utilities be connected to the Internet for operational effectiveness and to train staff on how to take advantage of this resource.

The following training sessions and resources are available for utilities:

- Water sector ICS training: hour-long webinars in ICS 100(b) basic training, ICS 200(b) initial action incidents, ICS 300 intermediate, ICS 400 advanced (training.fema.gov/emiweb/is/icsresource/trainingmaterials.htm)
- IS700(a) NIMS, IS800(b) national response framework
- Training materials (developed to supplement FEMA's public works training to provide more water-specific training)

### Training Methods

Table 6-1 provides a comprehensive summary of common training methods based on specific parameters. One training method does not fit all utilities. For example, a small and rural utility may find correspondence and online courses to be the most practical, while a large, metropolitan water utility has the resources for in-house training or partnering with other local utilities to teach similar subjects. The table can be revised to meet a utility's needs and experience. A training point to keep in mind is to use ICS and NIMS terminology in emergency preparedness documents to provide a seamless transition when an emergency expands from a local response to a higher authority, such as when a hurricane disaster goes from having a local impact to having a regional impact.

## CONDUCTING EXERCISES

A comprehensive exercise program is a critical element of any emergency preparedness program. There are two basic types of exercise activities: discussion-based exercises and operations-based exercises (Figure 6-4).

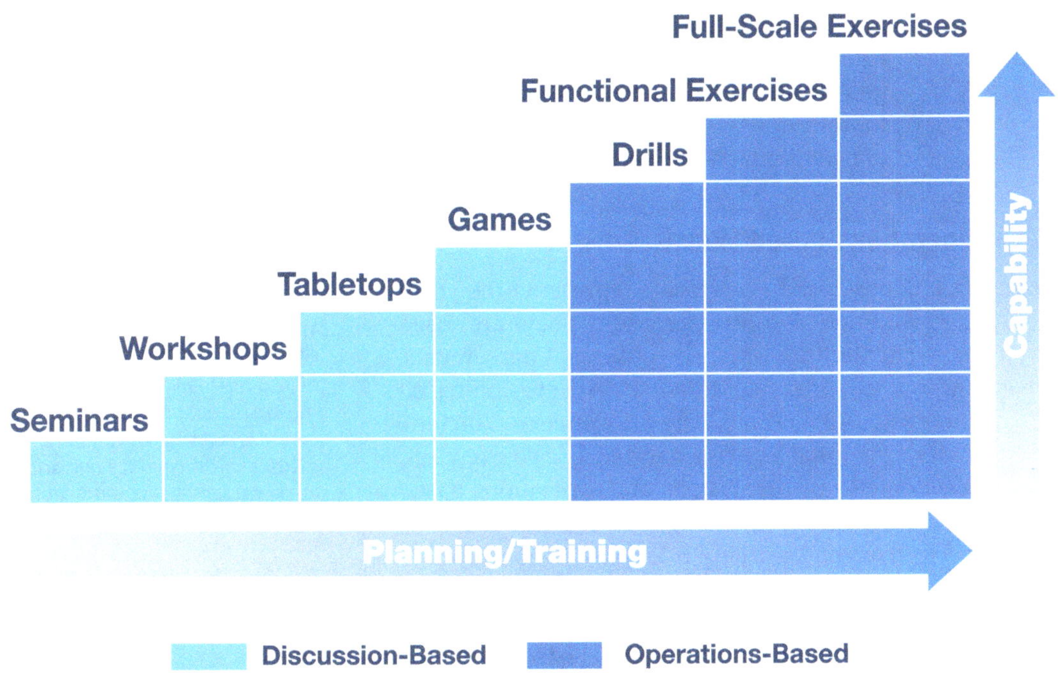


*Source:* DHS (2007)

Figure 6-4 Discussion-based and operations-based exercises

## Discussion-based Exercises

Discussion-based exercises are normally used as a starting point in the building-block approach of escalating exercise complexity. They include seminars, workshops, tabletop exercises, and games. These types of exercises typically highlight existing plans, policies, interagency/interjurisdictional agreements, and procedures. Discussion-based exercises are valuable tools for familiarizing agencies and personnel with the utility's current or expected capabilities.

These exercises typically focus on strategic, policy-oriented issues. Facilitators and/or presenters usually lead the discussion, keeping participants on track toward meeting exercise objectives. Following is a description of the different types of discussion-based exercises:

- *Seminars* are informal discussions used to orient participants to, or provide an overview of, authorities, strategies, plans, policies, and procedures.
- *Workshops* involve more participant interaction, and the focus is on achieving or building a product such as a draft plan or policy. To be effective, workshops should be highly focused on a specific issue and the desired outcome or goal should be clearly defined.
- *Tabletop exercises* involve key personnel who discuss hypothetical scenarios in an informal setting. This type of exercise can be used to assess plans, policies, and procedures or to assess the systems needed to guide the prevention of, response to, and recovery from a defined incident. Most tabletop exercises are aimed at facilitating an understanding of concepts, identifying strengths and shortfalls, and achieving changes in the approach to a particular situation. Participants are encouraged to discuss issues in depth and to reach decisions through slow-paced problem solving, rather than the rapid, spontaneous decision making that occurs under actual or simulated emergency conditions.

- A *game* is a simulation of operations that often involves two or more teams and uses rules, data, and procedures to depict an actual or assumed real-life situation. The goal of a game is to explore decision-making processes and the consequences of those decisions. A game does not require use of actual resources, and the sequence of events affects and is, in turn, affected by decisions made by players.

## Operations-based Exercises

> *ANSI/AWWA G440, Sec. 4.7.3: The utility shall conduct an exercise of their emergency preparedness plan annually. At least once every three years, the utility should conduct operations-based exercises involving outside response agencies. Tools and guidance exist on how to conduct exercises, including the Homeland Security Exercise and Evaluation Program (HSEEP) and EPA Toolbox.*

Operations-based exercises represent the next level of intensity for exercises. They are used to validate the plans, policies, agreements, and procedures solidified in discussion-based exercises. Operations-based exercises include drills, functional exercises, and full-scale exercises. An example of discussion- and operations-based exercises can be seen in Figure 6.4. These exercises are focused on clarifying roles and responsibilities, identifying gaps in resources needed to implement plans and procedures, and improving individual and team performance. Operations-based exercises are characterized by actual reaction to simulated intelligence; response to emergency conditions; mobilization of apparatus, resources, and/or networks; and commitment of personnel, usually over an extended period of time. The following define the operations-based exercises:

- A *drill* is a coordinated and supervised activity usually used to validate a single, specific operation or function in a single utility or organizational entity. Drills are commonly used to provide training on new equipment, develop or validate new policies or procedures, or practice and maintain current skills. A drill typically has a narrow focus performed in isolation and is measured against established standards in a realistic environment with immediate feedback.
- A *functional exercise* is designed to validate and evaluate individual capabilities, multiple functions, activities within a function, or interdependent groups of functions. Events are projected through an exercise scenario with event updates that drive activity at the management level. A functional exercise simulates the reality of operations in a functional area by presenting complex and realistic problems that require rapid and effective responses by trained personnel in a highly stressful, time-constrained environment.
- The *full-scale exercise* is the most complex type of exercise. The exercise is typically multiagency, multijurisdictional, multiorganizational, and validates many facets of preparedness. It focuses on implementing and analyzing the plans, policies, procedures, and cooperative agreements developed in discussion-based exercises and honed in previous, smaller, operations-based exercises. During full-scale exercises, events are projected through a scripted exercise scenario with built-in flexibility to allow updates to drive activity. It is conducted in real time, creating a stressful, time-constrained environment that closely mirrors real events.

## Exercise Design

Successfully conducting an exercise involves considerable coordination among participating agencies and officials. Exercises should be conducted in accordance with the US Department of Homeland Security's (2006) Homeland Security Exercise and Evaluation

Program (HSEEP), which is the national standard for all exercises. HSEEP is a capability- and performance-based exercise program that provides a standardized methodology and terminology for exercise design, development, conduct, evaluation, and improvement planning. FEMA (2013) provides a website with resources, guidance, and training for conducting exercises based on HSEEP. The USEPA has prepared an electronic tabletop exercise tool (consistent with HSEEP) to assist utilities in preparing and conducting a tabletop exercise. The Tabletop Exercise Tool for Water Systems: Emergency Preparedness, Response, and Climate Resiliency, contains 15 scenarios and many templates of exercise documents (USEPA 2010).

*The Homeland Security Exercise and Evaluation Program (HSEEP) is a capabilities and performance-based exercise program that provides a standardized methodology and terminology for exercise design, development, conduct, evaluation, and improvement planning.*

Exercise planning timelines establish target timeframes for attaining significant, exercise-related milestones, such as planning conferences, training, exercise conduct, after-action reporting, and improvement planning. These timelines should be consistent with the scheduling component of the entity's multiyear training and exercise plan. Timelines vary based on exercise scope and complexity. For example, exercise planners generally use longer timelines for tabletop exercises than for workshops and seminars and even longer timelines for complex or multijurisdictional full-scale exercises.

**Exercise planning team.** The best exercises are the most realistic and challenging. This can be accomplished by organizing a team of experienced personnel from multiple disciplines (e.g., operations, engineering, distribution, law, fire, public health) to help create objectives, scenarios, messages, and other materials. This exercise planning team is responsible for the successful execution of all aspects of an exercise, including exercise planning, conduct, and evaluation. The planning team determines exercise objectives, tailors the scenario to the entity's needs, and develops documents used in exercise simulation, control, and evaluation. While each exercise has its own planning team, personnel may carry over from one exercise to the next, and entities may find it advantageous to include team members with previous exercise planning experience.

Exercise scope, type, and complexity determine the number of exercise planning conferences necessary to successfully carry out the planning process. These may include a concept and objectives meeting, an initial planning conference, a midterm planning conference, and a final planning conference.

**Design and development.** The design and development process focuses on identifying objectives, designing the scenario, creating documentation, coordinating logistics, planning exercise conduct, and selecting a focus for evaluation and improvement planning. The exercise planning team should consider all the capabilities being evaluated. Each capability has specific tasks associated with it that should be performed and validated during the exercise. These capabilities and tasks should be used to formulate exercise objectives that reflect the exercising entity's specific needs, environment, plans, and procedures.

Well-defined objectives provide a framework for scenario development, guide individual organization's objective development, inform exercise evaluation criteria, and synchronize various agencies' efforts toward common goals to prevent duplication of effort and focus support on exercise priorities. The validation of capabilities is often accomplished by exercising and evaluating the agency plans or procedures that relate to the performance of the identified capabilities and tasks.

A scenario provides the storyline that drives an exercise. The first step in designing a scenario is determining the type of threat/hazard (e.g., chemical, explosive, cyber, natural disaster) to be used in an exercise. The hazards selected for an exercise should realistically stress the resources an entity is attempting to improve through its exercise program. The scenario should also be a realistic representation of potential threats and hazards faced by the exercising entity.

The next step in designing a scenario is to determine the site (i.e., facility) where exercise play will take place. Venue selection should reflect the hazard selected, allowing for realistic, exercise-based simulation of the hazard. Some important documents that may be needed as part of an exercise program include a situation manual, an exercise plan, a controller and evaluator handbook, a master scenario events list, a player handbook, an exercise evaluation guide, and exercise policies.

Logistical details are important aspects of an exercise. They can make the difference between a smooth, seamless exercise and one that is confusing and ineffective. Discussion-based exercises require attention to logistical details, such as the availability of appropriately sized and comfortable meeting and briefing rooms, food and refreshments, audiovisual equipment, facilitation and note-taking supplies, badges and table tents, registration assistance, and direction signs. Operations-based exercises require badge/role identification, access to restrooms, available food and water, onsite communications, arrangement of videotaping, props, site security, adherence to the weapons-check policy, and observation of safety precautions.

**Exercise conduct.** After design and development activities are complete, the exercise takes place. Important steps in exercise conduct include setup; briefings; management of facilitators, controllers, evaluators, players, and actors; and wrap-up activities.

For a discussion-based exercise, room layout is particularly important. When setting up an operations-based exercise, planners should consider the assembly area, response route, response operations area, parking, registration, observer/media accommodations, and a possible simulation cell facility. Restrooms and water should be available to all participants, observers, and actors. All individuals permitted at the exercise site should wear some form of identification. Perimeter security and site safety during setup and conduct are essential.

Presentations and briefings are important tools for delivering necessary exercise-related information to participants. A discussion-based exercise generally includes a multimedia presentation to present the scenario and accompany the situation manual. An operations-based exercise may include briefings for controllers/evaluators, actors, players, and observers/media. Briefings and presentations are opportune times to distribute exercise documentation, provide necessary instructions and administrative information, and answer any outstanding questions.

In both discussion- and operations-based exercises, facilitators and controllers guide exercise play. During a discussion-based exercise, the facilitator is responsible for ensuring that participant discussions remain focused on the exercise objectives and ensuring all issues and objectives are explored as thoroughly as possible within the available time. In an operations-based exercise, controllers plan and manage exercise play, set up and operate the exercise incident site, give key data to players, and may prompt or initiate certain player actions.

Evaluators are selected from participating entities to evaluate and comment on designated functional areas of the exercise and are chosen based on their expertise in the functional areas they evaluate. Evaluators have a passive role in the exercise and should only record the actions/decisions of players; they should not interfere with exercise flow. Evaluators use exercise evaluation guides to record observations and notes.

Players have an active role in responding to a simulated incident by either discussing (in a discussion-based exercise) or performing (in an operations-based exercise) their regular roles and responsibilities. Actors are volunteers who simulate specific roles, such as disaster casualty victims, to add realism to an exercise. Simulators perform the roles of individuals, agencies, or organizations that are not actually participating in the exercise to drive realistic exercise play.

> *ANSI/AWWA G440, Sec. 4.7.4: The utility shall conduct a critique following any exercise or incident documenting strengths, areas for improvement, and lessons learned. Common formats include an After Action Report and an Improvement Plan. Both forms are available through FEMA. Any relevant findings shall be incorporated in the emergency plan.*

**Evaluation.** Exercise evaluations benefit a utility's preparedness by identifying and documenting strengths and areas for improvement. These evaluations feed improvement planning activities and enable utilities to make real progress in their level of preparedness. Evaluation takes place using predeveloped exercise evaluation guides such as the standardized exercise evaluation guides provided in HSEEP.

The evaluation process for all exercises includes a formal exercise evaluation, integrated analysis, and drafting of an after-action report/improvement plan (AAR/IP). Both hot washes (for exercise players) and debriefs (for facilitators or controllers and evaluators) follow discussion- and operations-based exercises.

An AAR/IP is used to provide feedback to participating entities on their performance during the exercise. The AAR/IP includes a summary of exercise events and analysis of performance of the tasks identified as important during the planning process. It also includes an evaluation of the achievement of the selected exercise objectives and demonstration of the overall capabilities being validated. The IP portion of the AAR/IP includes corrective actions for improvement, along with timelines for their implementation and assignment to responsible parties.

**Improvement planning.** During improvement planning, corrective actions from the AAR/IP, such as additional training, planning, and/or equipment acquisition, are assigned, with due dates, to responsible parties. They are then tracked to completion, ensuring that exercises result in tangible benefits to preparedness (Figure 6-5).

*Source:* DHS (2007)

**Figure 6-5 Exercise program management**

The IP portion of an AAR/IP converts lessons learned from the exercise into concrete, measurable steps that result in improved response capabilities. It specifically details the actions that the participating entity will take to address each recommendation presented in the draft AAR/IP, who or what agency will be responsible for taking the action, and the timeline for completion.

Once recommendations, corrective actions, responsibilities, and due dates are clearly identified in the IP, the exercising entity ensures that each corrective action is tracked to completion. Exercising entities review all exercise evaluation feedback and resulting IPs to assess progress on enhancing preparedness. This analysis and information are incorporated into the capabilities-based planning process because it may identify needs for additional equipment, training, exercises, and coordination.

Continual IP tracking and implementation should be part of a corrective action program within each participating entity. A corrective action program ensures IPs are "living, breathing" documents that are continually monitored and implemented and that they are part of the larger cycle of improving preparedness. The USEPA has prepared a factsheet that highlights lessons learned from water utilities that conducted state-level water sector emergency response exercises from 2009 to 2011.

## DOCUMENTATION

As with all training, course and exercise attendance records should be maintained, as well as an outline of the course and any test results. With exercises, the sign-in sheet, copy of the exercise, AAR/IP, and any evaluations need to be maintained. A separate listing of employees and their emergency capabilities (e.g., ICS, first-aid trained, trained in cardiopulmonary resuscitation) should also be maintained as metrics that can be used to demonstrate the utility's commitment to preparedness.

## CONCLUSIONS

Training and exercise of an ERP are very important to providing an effective response during an actual emergency. It is recommended that utilities adopt a multiyear training and exercise program. Training is an effective way to introduce plans and procedures to the employees. Utilities should carefully choose from the many methods that are available for training their employees. Exercise, on the other hand, involves simulating an emergency in various forms. When pursuing an exercise program, it is important to start with less complex forms of exercises and gradually build up to greater complexity.

## REFERENCES AND SUPPLEMENTAL READINGS

Department of Homeland Security. 2006. *Homeland Security Exercise and Evaluation Program (HSEEP): Volume I: HSEEP Overview and Exercise Program Management.* www.michigan.gov/documents/deq/deq-wb-wws-HSEEP_Vol_I_271850_7.pdf.

FEMA (Federal Emergency Management Agency). 2013. Homeland Security Exercise and Evaluation Program. www.fema.gov/hseep.

USEPA (US Environmental Protection Agency). 2011. *How to Develop a Multi-Year Training & Exercise (T&E) Plan: A Tool for the Water Sector.* water.epa.gov/infrastructure/watersecurity/emerplan/upload/epa816k11003.pdf.

USEPA. 2010. Tabletop Exercise Tool for Water Systems: Emergency Preparedness, Response, and Climate Resiliency (TTX Tool) Overview. EPA-C-10-001. water.epa.gov/infrastructure/watersecurity/techtools/ttx.cfm.

# Chapter 7

# Mitigation

To increase its resilience and preparedness, a utility must actively manage risk. Ways to identify and reduce risk are presented in Chapter 2; this can also be done through ongoing discussions and activities with staff and outside partners (see Chapters 1 and 3–6). This chapter continues the description of how to effectively mitigate risk by using strategies to avoid risk (e.g., moving critical infrastructure to a safer location, adding redundancy to systems) and how to reduce the negative effects of risk (e.g., hardening buildings, using alternate water sources). Information about mitigating the consequences of remaining risks through preparedness planning and consequence management is found in Chapter 3. This chapter also presents hazard mitigation planning, which is a legal plan that provides eligibility to apply for and receive pre- and post-disaster grant funding. It includes an outline of a hazard mitigation plan, planning resources, and examples of specific measures that can be taken to address risks that have the potential to impact a utility.

## WHAT IS MITIGATION?

As defined by Merriam-Webster, to mitigate something is to make it less severe, harmful, or painful. We take many routine actions in our everyday lives to mitigate the effects of identified risks. For example, every time you fasten a seatbelt, put on sunscreen during a hot summer day, or lock the doors of our cars, you are mitigating identified risks.

Mitigation is a key phase in the emergency management cycle that includes planning and procedures that can be applied to all aspects of a water utility's operations, resources, and assets. As it applies to effective emergency preparedness at a water utility, mitigation can be referred to in different ways. In an emergency response plan, it may be referred to as "mitigation," "risk mitigation," or "risk management." Within an emergency response plan, it is more specifically referred to as "hazard mitigation" when referring to actions taken for the mitigation of known natural hazards and as "risk mitigation" when referencing actions taken relative to hazards that may affect business continuity. All these terms are synonymous in that they describe steps aimed at lessening the effects of identified hazards or risks. However, it is best to use the terms in an emergency preparedness program as they are known relative to the hazards identified and the actions to be taken.

## HAZARD MITIGATION

Hazard mitigation is any sustained action taken to reduce or eliminate the long-term risk to life and property from hazard events. It is an ongoing process that occurs before, during, and after disasters and serves to break the cycle of damage and repair in hazardous areas. Effective hazard mitigation for any water utility relies on the geographically specific identification and assessment of naturally occurring hazards and risks followed by the strategic implementation of actions to be taken that best serve to minimize the potential effects of the hazard or risk.

### Hazard Mitigation Planning

As part of its emergency preparedness planning effort, a utility should prepare a local hazard mitigation plan. Hazard mitigation planning has legal basis under the Robert T. Stafford Relief and Emergency Assistance Act (Public Law 93-288 as amended, 42 U.S.C. 5125 et. Seq.). Under this law, the Federal Emergency Management Agency (FEMA) requires state, tribal, and local governments to develop and adopt hazard mitigation plans as a condition for receiving certain types of nonemergency disaster assistance, including funding for mitigation projects (see the FEMA Hazard Mitigation Assistance Grant Program section later in this chapter). A water utility's hazard mitigation plan is submitted to and approved by FEMA either as a single-jurisdiction, utility-specific plan or as part of a multijurisdictional regional planning effort. Once the plan is approved, the utility is eligible under the Stafford Act to apply for and receive pre- and post-disaster mitigation grant funding.

The core approach and components of the plan are consistent, whether the mitigation planning effort is for a single jurisdiction and is agency specific or whether it is part of a multijurisdictional planning effort (i.e., a utility might develop a utility-specific annex to be attached as part of a regional municipalities plan). Following are the steps involved in developing a mitigation plan:

1. Determine and identify the area covered in the plan. For water utilities, this will likely be specific to the geographic extent of its jurisdictional boundaries and/or system. It may also include neighboring watersheds that affect the utility's water supply portfolio. Once the area is determined, the plan will be focused on consideration of the hazards and risks associated with the specific geographic area.
2. Build the planning team. Under the strong leadership of a planning effort champion (in most cases, this effort is led by the utility's emergency preparedness officer), the water utility should identify those who are best suited for the planning effort. For a single-jurisdiction utility plan, the team should include representatives from management, engineers, and technical support staff. For a multijurisdictional effort, the team should include individuals with a diversity of subject matter experience, with the water utility representation being consistent with the other participating agencies.
3. Create an outreach strategy to inform the public served. Whether the planning effort is to serve a single jurisdiction or multiple agencies, there should be a plan to communicate and inform the public served by the plan that the effort is being undertaken. Informing the public of the plan and making them a recognized stakeholder in the effort is critical to building trust. This is possible through an

understanding of the hazards and risks that the utility faces. It also serves to provide the public with information on the mitigation strategies being pursued to minimize impact to the community.

4. Review and understand the agency's and community's capabilities to mitigate. Once the area is determined and the planning team is established, current and potential capabilities to effectively mitigate hazards should be assessed. The depth and breadth of capabilities to mitigate vary widely from area to area. This step is intended to identify and understand all capabilities so that mitigation, which is based on hazard assessment findings, can be realistically planned. As a starting point, it is important to understand what might be undertaken in a mitigation plan. Capabilities being considered for mitigation should include existing and future financial resources, benefits available through mutual agency efforts and coordination, benefits available through public outreach and education, and any existing insurance coverage.
5. Conduct a hazard/risk assessment. Every mitigation planning effort should be based on a comprehensive hazard/risk assessment. Chapter 2 provides a comprehensive explanation and approach for performing an effective risk and resilience assessment. As outlined in Chapter 2, the assessment should take all hazards and risks into consideration. For inclusion into an agency's local hazard mitigation plan, the hazards and risks are typically focused on naturally occurring events. Specific hazards that might be included are associated with earthquakes, landslides, wildfires, floods, and other weather-related events. The process of performing a risk and resilience assessment, as outlined in Chapter 2, is critical to the development of an effective hazard mitigation plan.
6. Develop a mitigation strategy. At its core, the most important component of a mitigation plan is the mitigation strategy. The strategy serves as the long-term plan for actions to be taken to reduce the impact of the risks and hazards determined by the assessment. The mitigation strategy, which outlines how the agency or community will accomplish the plan's mission, consists of three required core elements. First, the strategy must clearly outline the plan's goals. Specifically, the desired long-term outcomes of the plan need to be identified. Second, the specific actions that need to be taken to reduce the risks from hazards must be listed. Third, the action plan must identify priorities and desired schedules for the actions to be implemented. Development of a mitigation strategy requires great effort, and realistic objectives must be established. The mitigation strategy will serve as the road map for achieving a more resilient and risk-tolerant community.
7. Review and adopt the plan. Once the plan is developed, all key stakeholders should review it before adoption by the jurisdiction(s) included in the plan. Outreach and clear, transparent communication of the review and adoption process should be made to those who are served by the plan.
8. Keep the plan current. Much like a water utility's emergency response plan, the hazard mitigation plan is intended to be a "living" document that is regularly reviewed and updated. The plan must be kept current to be effective. When mitigation efforts are made to minimize the risk from any hazard, the plan should be updated to reflect this. Accordingly, if new risks or hazards are identified, the plan should be updated.

Many additional resources are available online to provide guidance in developing and updating a hazard mitigation plan. These sites also provide guidance and relative plan requirements necessary for states and local jurisdictions to apply for hazard mitigation grant funding programs. Federal agency-supported sites are listed in the References and Supplemental Readings section.

## FEMA HAZARD MITIGATION ASSISTANCE GRANT PROGRAM

The FEMA Hazard Mitigation Assistance (HMA) grant programs were developed to help reduce the risk to individuals, property, and utility infrastructure from naturally occurring hazards. The three core grant assistance programs that comprise the HMA include the following:

- Hazard Mitigation Grant Program (HMGP)—provides grant funding for hazard mitigation projects to be undertaken following a disaster when a presidential declaration of a major disaster has been issued.
- Pre-Disaster Mitigation Program—provides grant funding for hazard mitigation programs and projects that have identified to reduce overall risk to the population, structures, and critical infrastructure before the occurrence of disaster.
- Flood Mitigation Assistance Program—provides grant funding assistance aimed at mitigating the risk to flood-damaged properties so that future flooding events will not result in a recurrence of similar losses.

Additional information on the HMA programs can be found on the FEMA website (see the References and Supplemental Readings section).

FEMA's HMA grants are provided to eligible applicant states, tribes, and territories that, in turn, provide subgrants to local governments and jurisdictions, including water utilities. Each state has a dedicated program for the management of jurisdictional hazard mitigation planning and to provide assistance for those applying for hazard mitigation grant funding. Local governments and agency jurisdictions are encouraged to contact their state hazard mitigation officer for support and assistance. General information and current contact information for all states, tribes, and territories can be found on the FEMA website.

## HAZARD MITIGATION ACTIONS FOR UTILITIES

In performing a comprehensive risk assessment, as outlined in Chapter 2, all critical utility assets, components, and operations are identified. The assessment should determine and assign a level of criticality to each item as it relates to the utility's core mission. In addition, the asset's current condition, cost to replace or upgrade, and projected service life should also be determined through the assessment. The process of determining which assets the utility should identify for mitigation projects can be based on a priority ranking.

A utility can pursue many mitigation actions depending on the risks and hazards that have been identified. For example, the list of potential mitigation activities compiled by a utility located in an area subject to devastating earthquakes will be different from the list compiled by a utility located in an area subject to hurricanes. Ultimately, the utility should pursue mitigation actions that are possible to undertake and to complete and that will yield the greatest value for the investment.

The following sections provide examples of mitigation actions a utility might consider based on the information gleaned from the hazard/risk assessment. The mitigation actions outlined here are provided as they relate to specific assets typically found at a water utility.

Personnel

- Education and training on emergency preparedness and response
    - Importance of personal/family emergency preparedness planning
- Safety training
- Continuity of operations planning actions

Source water supplies and transmission

- Watershed protection and monitoring
- Source water conveyances
  - Open channel
  - Closed pipelines and aqueducts
- Source water storage facilities
  - Lakes and open-water reservoirs

Treatment and production facilities

- Piping systems
- Basins and related mechanical systems
- Chemical storage and piping systems

Distribution system components

- Pipelines
  - Transmission pipelines (12 in. and larger)
  - Distribution service pipelines
- Tanks and reservoirs
- Pump stations and regulators

Control systems

- Supervisory control and data acquisition systems

Communication systems

- Radio systems
  - Voice communication systems
  - Data communication systems

Emergency operations centers (EOCs)

- Primary EOC
  - Dedicated outfitted and ready
  - Ready for stand-up as needed
- Secondary EOC
  - Ready for stand-up as needed

## Personnel

For any water utility, the key system component is personnel. Personnel safety is the utility's first responsibility. There are four ways to help mitigate personnel shortages:

- Education
- Cross-training
- Replacement
- Personnel support
- Safe workplace

Training and practice based on a completed emergency preparedness plan are covered in Chapter 6.

**Education.** Education is the first step in ensuring employee safety or, more precisely, in minimizing casualties during an emergency. Employees should review the outcomes of the hazard summary and risk analysis. Both management and staff must understand that personnel safety is the first concern in a disaster. Staff should never risk their lives during an emergency. For example, an improperly protected worker should never attempt to rescue a coworker who is overcome by a chlorine leakthat work could be overcome by the chlorine leak, too.

Education for both staff and management is a continuous process. Employee input should be sought when determining mitigation actions because they are most familiar with system components and processes.

Extend education beyond the workplace as well. Teach employees what to do at home in case of a disaster. Family safety plans help ensure quicker return of off-duty employees.

**Cross-training.** Many critical system components are operated or maintained by only one or two key people. Periodic employee cross-training can ensure backup capability for critical tasks and processes. Another example of cross-training is preparation of administrative and laboratory personnel to assist field crews during an emergency.

**Replacement.** A mutual aid agreement can be established with other utilities to supply replacement personnel. Participating utilities must provide their mutual aid partners with a detailed description of the system and its operations as well as a copy of the emergency plan. Joint training exercises should be considered.

Private contractors are another source of replacement personnel. A contractual agreement that includes hourly rates and specifies the types of services that can be provided should be established. The contractor should be provided with a copy of the emergency plan. The agreements need to be updated regularly to account for changes in services and rates and for any modifications made to the emergency plans.

**Personnel support.** One key mitigation measure that is often overlooked is support of personnel during emergency responses. Support includes food, drinking water, first-aid and medical care, sleeping facilities, and sanitary facilities. Contractual arrangements can be established with a local food catering service. While such an arrangement may not be effective during an area-wide disaster (e.g., an earthquake), it can prove useful during more limited emergencies. Emergency rations such as military meals-ready-to-eat and drinking water should be stocked and replaced or replenished periodically.

A well-prepared water utility should contact local vendors for portable toilets and shower facilities. Emergency blankets and miscellaneous sundries (e.g., snack foods, soap, dental needs, and over-the-counter medicines such as aspirin and antidiarrheals) should also be stocked.

**Safe workplace.** Utilities must provide a workplace that is safe for employees during normal operations and as safe as possible in emergencies. This includes meeting applicable US Environmental Protection Agency and Occupational Safety and Health Administration regulations and other federal, state or provincial, and local regulations.

## Source Water Supplies and Transmission

Mitigation measures that relate to source water and transmission range from providing alternate sources and protecting wellheads to retrofitting dams or aqueducts. The following section provides mitigation actions for surface water and groundwater sources.

**Surface water sources.** Surface water sources include watersheds and surface sources such as rivers, reservoirs, and dams. Large-scale watershed damage from hurricanes, landslides, fires, flooding, and volcanic eruptions is difficult to mitigate. For example, stabilizing slopes vulnerable to landslides could be a Herculean task. The effort could damage the watershed, and the slopes are often best left undisturbed. However, these critical areas should be monitored to determine whether conditions exist that could contribute to or exacerbate disaster hazards. Such conditions include illegal dumps, a buildup of combustible material, hazardous material spills, and construction activities that cause erosion. Determine which agency is responsible for fire suppression in the watershed and coordinate a fire suppression plan. The plan should include initial fire notification, discussion on the use of fire retardants, and similar actions.

Mitigation actions for watershed damage or widespread contamination include providing automated monitoring equipment, using alternate sources or intakes, and modifying source water treatment at the plant (e.g., expanding settling basin capability). Mitigation of smaller-scale watershed or surface water source contamination from hazardous material spills is best handled using an effective emergency spill response plan. This type of hazard is noted in Chapter 2, and key elements of this type of hazard-specific plan include notification, containment, and cleanup.

The effects of natural hazards on open reservoirs can be mitigated by controlling access, identifying alternate sources, providing alternate intake locations, and providing flexible treatment facilities. Access can be controlled by installing fences, gates, and signs; closing unnecessary roads; and increasing security patrols.

Dams must be closely inspected because of the potential damage and loss of life posed by failure. Retrofit measures include raising the dam's height to accommodate higher design floods and seiches, increasing spillway capacity to provide for design floods, and improving structural strength to withstand seismic forces. As an example, in 1988, the Bear Valley Dam in southern California was converted from a concrete multiple-arch dam to a gravity dam. Earlier studies revealed that an arch dam could be overstressed by seismic design loads for the area. During the 1992 Landers and Big Bear earthquakes (7.2 and 6.7 on the Richter scale, respectively), the retrofitted structure survived 15 seconds of horizontal forces that were greater than half the structure's weight (Denning 1993). Another example is the East Bay Municipal Utility District's (California) rehabilitation of earthquake-prone hydraulic fill dams during the 1970s.

Impact to water service from intake structure damage can be mitigated by providing temporary bypass pumps, having filtration materials (e.g., boom floats, screens) on hand, providing multiple intakes, strengthening the structures to resist wind and earthquake hazards, and ensuring that access to the structures is available by alternate means such as boats. Valves or other methods to drain or shut off flow to the system should be redundant (i.e., failure of any one component will not put the entire facility out of service), hazard resistant, and properly maintained.

One important alternate source of emergency water supply is an interconnection with an adjacent water system. For more information, refer to the section on mitigating disaster effects on distribution components later in this chapter.

**Groundwater and wells.** Methods of mitigating groundwater contamination include identifying alternate water sources, providing shutoff valves, maintaining adequate setbacks from sewage pipes and disposal systems, and installing wellhead protection and well seals. A spill response plan and hazardous waste disposal plan are also recommended to mitigate groundwater contamination. Note that although earthquakes can produce changes in hydrogeology, deeper aquifers are less susceptible to this problem.

Many natural disaster hazards can cause power outages to the well pumps and other system components. Water utilities should request that electric power to well pumps receive a high priority for service and provide appropriate information to their electric utility. Equipping pump stations with backup electrical power generators and adequate fuel and providing connections for portable generators at wellheads are prudent mitigation measures. Refer to the following section on treatment for additional electrical power mitigation actions.

Mitigation actions for the pump station structure and equipment are the same for similar system components such as treatment plants. These actions are discussed in the following section on treatment.

Wells are vulnerable to earthquake hazards. Mitigation actions can be applied to the following subcomponents:

- *Casing:* Wells should not be installed in liquefiable soils or soils subject to lateral spreading that could bend the casing. If such soils are unavoidable, mitigation actions include the following:
  - stabilize the liquefiable layer,
  - use double well casing, or
  - use a submersible pump (straight drive shaft not required).
- *Well screens and slotted casing:* Provide a good-quality well screen to avoid sanding problems with slotted casings.
- *Connecting pipes:* Provide flexible piping to mitigate settlement or relative movement hazards.
- *Well pump materials:* Use ductile-iron or steel castings to prevent pump and motor castings from breaking during ground shaking.
- *Electric power:* Anchor transformers, controls, and standby generators.

## Water Treatment

Treatment facilities contain several highly complex components that are critical to system operation. A redundant-component philosophy should be used during the design phase of water-treatment facilities.

Under a worst-case scenario, the treatment plant will be offline. In some cases, it may be necessary to isolate the facility from the system (e.g., cases of contamination at the plant). Confirm that isolation valves are installed and that their locations are well marked. Mitigation for the off-line scenario includes providing alternate treatment such as increasing production at another facility in the system or identifying alternate sources of treated water from adjacent systems. Another measure is to provide a bypass with disinfection capability.

Treatment facilities should be fire resistant, and all appropriate fire prevention measures should be implemented. Contact the local fire department for support and request regular facility inspections.

Treatment facilities are often located near flood plains or in flood-prone areas. Facility design should include hydrologic and hydraulic analyses.

Mitigation of potential hazard effects for treatment components is discussed in the following sections. Additionally, refer to AWWA and ASCE's *Water Treatment Plant Design* (2012) for a detailed discussion of the behavior and design of treatment plants that are subject to earthquake hazards.

**Facility structures.** Treatment plant structures and other water system structures such as pump stations are often single-story, symmetrical, shear-wall buildings that perform well in earthquakes, high winds, and other disasters. Design considerations and mitigation actions include the following:

- Earthquake
  - Use reinforced masonry and other hazard-resistant construction.
  - Avoid liquefiable soils for structure foundations.
  - Avoid using dissimilar foundation types when adding onto a structure.

- Earthquake and wind
  - Ensure roof and wall connections and roof and foundation connections are adequate.
  - Correct modifications that may have weakened structures such as removed bracing or bearing walls cut for doors.
- Vandalism
  - Install adequate locks, window security, and lighting.
  - Install intrusion-prevention devices such as electronic keys, identification-card checkers, 10-key code units, joint 10-card/10-key checkers, and noncontact-type identification-card checkers to control access to the strategic facilities of the system (AwwaRF and JWWA 1993).
  - Install situation display monitors that consist of closed-circuit television systems with high-sensitivity image pickup tubes, including camera and sound collectors, as well as alarm systems with ultrasonic, heat, or beam sensors and magnetic switches to detect intruders (AwwaRF and JWWA 1993).

**Electrical power and instrumentation.** Standby power should be available for critical components such as pumps and controls. This can be done by establishing a duplicate power supply from a separate power distribution point and following different routing. Specific mitigation actions for electrical power and instrumentation components and backup power include the following:

- *Transformers:* Anchor ground-level units to foundation pads and protect them from flooding. Securely anchor pole-mounted transformers to the pole and ensure that the pole is secure in the ground. Transformer aerial conductors must be protected from swinging shorts that occur during earthquakes and high winds.
- *Emergency generators:* Depending on expected seismic acceleration, generators may be anchored directly to the floor or anchored via snubbers on vibration-isolated bases. Figure 7-1 shows a collapsed spring-vibration isolator. Store and maintain

*Source:* D.B. Ballantyne

**Figure 7-1 Collapsed spring-vibration isolator supporting an emergency generator in Whittier, California**

portable generators at various locations throughout the distribution system to avoid a complete loss of standby power. All emergency generators must be regularly inspected and tested while under load. Ensure that all emergency generators have sufficient fuel, appropriate cooling, fully charged starting batteries, and secure exhaust routing.

- *Electrical cabinets:* Anchor electrical cabinets to the floor and restrain the top of the cabinet using angle clips or another secured structural element.
- *Motors:* Provide automatic shutdown capability to avoid voltage or phase fluctuation that may cause damage.
- *Batteries:* Batteries for backup instrumentation power or starting generators should be firmly anchored to prevent toppling. They should be raised off the floor to prevent potential water damage. All batteries should be inspected and serviced at regular intervals.
- *Telemetry:* A radio system with adequate backup power can provide a replacement for broken cables or downed phone lines. Lightning-strike protection should be provided if needed.
- *Computers:* Computers and related networking systems can be located on special base isolation units that can resist seismic forces. Desktop computers should be secured to desks with commercially available restraints. Be sure to back up computer files and store them at separate locations.

The effects of computer failure can be minimized by providing the following elements:

- offsite backup for the computer system,
- preventive maintenance program for hardware,
- uninterruptible computer power supply,
- adequate computer room security,
- compatible computer hardware components,
- written procedures for testing software and addressing operational problems,
- adequate training for programmers,
- control of operating files and documentation by data processing personnel,
- virus prevention,
- adequate system and network security,
- lightning-strike protection, and
- anchored, suspended ceilings.

**Equipment, chemical storage, and piping.** Most equipment and chemical storage containers in a treatment plant will be damaged from sliding and toppling. Simply anchoring or restraining equipment and containers can prevent much of the damage. Equipment supported on legs should have angle braces installed on the legs for additional strength. Waterproof plastic sheeting should be stockpiled in case equipment or chemicals become exposed to the elements. An adequate supply of critical spare parts and tools should be maintained in several secure storage locations if possible.

Piping damage due to sliding or toppling equipment can also be reduced by anchoring and using flexible connections. Some specific subcomponents and mitigation actions are described below.

*Chlorine and chemical containers*. Chlorine cylinders (150 lb [70 kg]) should be stored vertically and restrained at the top and bottom. One-ton (1,000-kg) containers should be stored horizontally and secured in cradles or anchored with chain binders or nylon straps. Check with tank and container manufacturers for the proper method of restraint.

Chlorine systems can be designed and constructed with numerous automatic control systems. Such systems can indicate the extent and location of a leak, actuate chlorine scrubbers, close valves, shut down equipment, and isolate affected areas. Consider using sodium hypochlorite as an alternative to chlorine.

*Buried tanks.* Tanks that are subject to sinking or floating in liquefiable soils or during flooding can be restrained to reduce damage. Soils can also be stabilized.

*Piping.* To prevent damage from swinging pipe that is supported on rods from ceilings, support the piping in three directions at right angles. Do the same for chlorine tubing. Pipe appurtenances such as valves that rise vertically from a pipe can act as pendulums that amplify ground motion. These should be appropriately anchored. Flexible connections should be provided for buried piping that is located at a building's edge to prevent differential settlement and differential lateral movement.

*Laboratory equipment and chemicals.* Shallow curbs or a bar across the face of the shelves can be installed to prevent laboratory equipment and chemicals from falling.

**Process basins and tanks.** Mitigation of damage to process tanks is best handled during the design phase. Retrofitting tanks is difficult and often involves substantial and expensive reconstruction. The design of treatment facility tanks should include flexibility and redundancy. For example, there should be multiple connections between tanks to provide alternative piping configurations for tanks in series.

Specific earthquake design considerations for process tanks include the following:

- Provide breakaway design or other appropriate methods for submerged elements or baffles to allow easy replacement;
- Use appropriate codes and design methods for concrete tanks,
- Perform a geotechnical investigation, stabilize the soil as necessary, and use the appropriate foundation; and
- Design with neutral buoyancy, keep the tank full, or use piles to hold the tank in place.

**Storage tanks.** As with process basins, the best time to implement disaster mitigation features for storage tanks is during the design phase. Refer to the appropriate tank standards for more information (ANSI/AWWA D100, ANSI/AWWA D110). Design considerations include the following:

- Use appropriate design values for earthquake, wind, and other live loads;
- Perform geotechnical, geological, and seismological studies to evaluate foundation designs;
- Provide adequate freeboard baffling and strength to minimize sloshing;
- Provide correctly positioned isolation valves;
- Allow pipe flexibility;
- Make provisions for security, such as fences, gates, lighting, alarms, and locked valves and hatches; and
- Ensure access is available in case of flooding or other disaster hazards.

Existing tanks should be evaluated for structural soundness and performance during disasters. Specific mitigation actions for existing tanks of various types include the following:

- *Wire-wrapped concrete tanks*: Check for vertical cracking, delamination, and water corrosion staining on the outside of the tank. Check for installation of seismic cables.

- *Steel standpipes and ground storage tanks:* Install or upgrade anchors or stiffeners (Figure 7-2). Upgrade the foundation if it is inadequate. Check the capability of walls to handle additional stress.
- *Horizontal tanks:* Upgrade support to carry loads in three directions at right angles.
- *Elevated steel tanks:* Upgrade with additional foundation strength, base isolation, and lateral support. Upgrade the steel structures and allow the bracing to yield before a connection failure.

## Distribution Systems

Owing to their large numbers and widespread dispersal, mitigating disaster effects on distribution system components can be difficult. One key to locating and assessing any impact to the system is an accurate and up-to-date system plan, which should be maintained in numerous locations.

Mitigation measures for pumping stations are discussed in the source water and transmission section. Computers may be useful for gathering and managing regional network information when determining problem areas and damage. They can also be effective in monitoring and controlling emergency control systems and in maintaining equipment and supply inventories.

**Pipelines, valves, and other appurtenances.** Design methods and types of material can mitigate disaster effects on pipelines, valves, and other appurtenances.

**Pipeline design, construction, and upgrading.** As with other distribution system components, mitigating hazards for pipelines is best accomplished during the design and construction phase. Refer to the appropriate AWWA pipe standard for guidance. Some general considerations include the following:

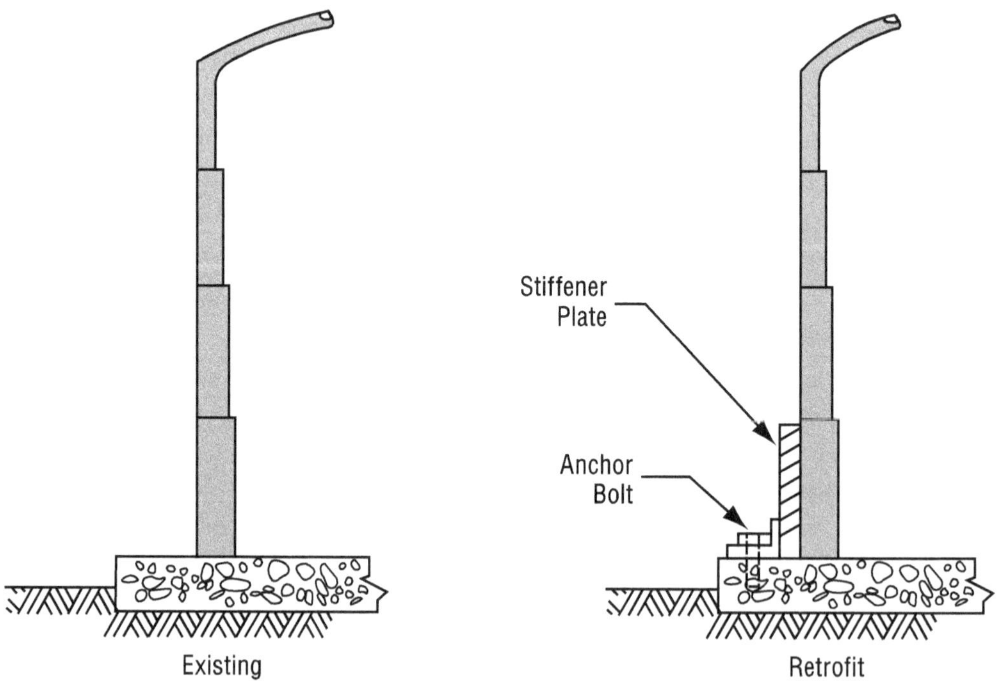


*Source:* American Society of Civil Engineers, Technical Council on Lifeline Earthquake Engineering (1983)

Figure 7-2 Mitigation option for water storage tanks

- Use the most current pipe material available (pipe materials are continually being improved);
- Avoid installing pipe in liquefiable soils and areas prone to flooding and landslides;
- Use pipe materials suitable for the soil conditions and expected hazard;
- Route pipelines at right angles when crossing known faults;
- Use flexible expansion joints at fault crossings and where soils change to liquefiable soils;
- Use compressible backfill material and shallow cover when crossing known faults to permit pipe/soil movement;
- Stabilize soils with vibroflotation, stone columns, grouting, compaction, or other methods;
- Use pile supports where required;
- Provide operational flexibility and redundancy;
- Build in emergency response capability such as isolation valves, check valves, and automatic shutoff valves; and
- Maintain a rigorous cross-connection control program.

Upgrading of existing pipelines should begin with the critical segments identified in the vulnerability assessment. When upgrading pipeline segments, use the design considerations listed above for new pipelines. See Figure 7-3 for an example of a damage mitigation procedure.

*Performance of pipe materials.* The performance of pipeline materials and joints during earthquakes has been documented in many studies. The following summarizes some key results of these studies.

In general, steel, ductile-iron, and polyethylene pipe successfully accommodate more ground deformation than cast-iron or asbestos-cement pipe. Restrained joints provide continuity to pipelines and can withstand moderate ground movement. Bell-and-spigot–style joints allow flexibility but can easily be pulled apart.

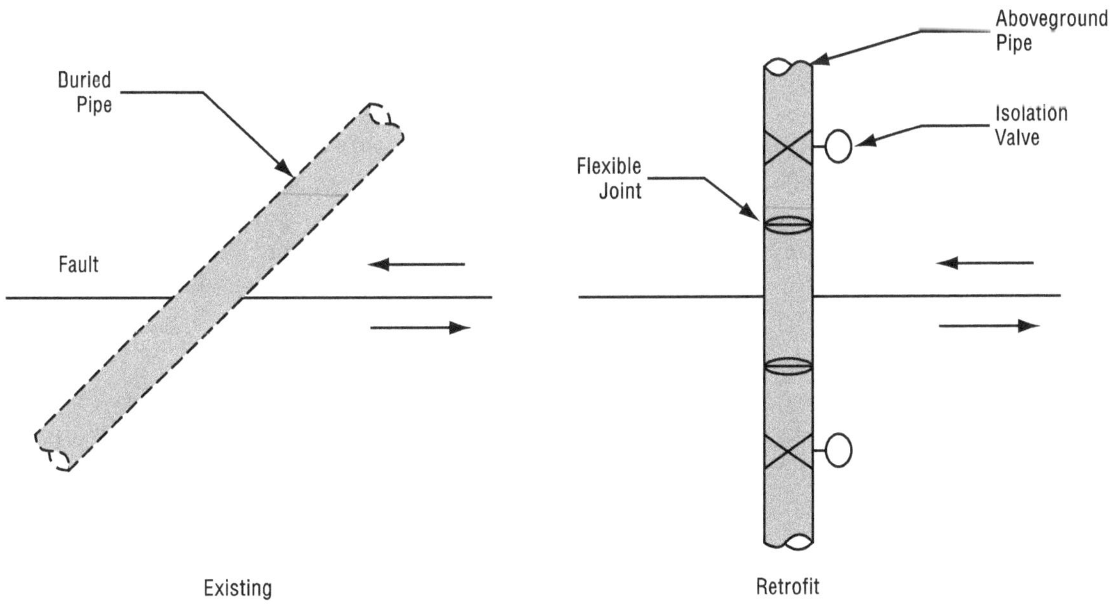


*Source:* American Society of Civil Engineers, Technical Council on Lifeline Earthquake Engineering (1983)

**Figure 7-3 Mitigation options for pipelines at fault crossings**

High-performance pipelines, such as electric arc-welded steel, restrained joint segmented steel, polyethylene, and restrained joint ductile-iron, can accommodate permanent ground deformation up to approximately 20 in. (500 mm). Average- to high-strength materials, such as restrained joint polyvinyl chloride (PVC) or restrained joint concrete cylinder pipe, can handle approximately 4 in. (100 mm) of permanent ground deformation. Average-strength pipelines, such as unrestrained joint ductile-iron, PVC, concrete cylinder, reinforced concrete, cement, asbestos-cement, and segmented steel pipe, can accommodate ground deformation of less than 1.5 in. (40 mm) or up to the bell-and-spigot insertion length allowance. Average- to low-strength pipelines (tolerating less than 0.75 in. [20 mm] of ground deformation) include cast iron, vitrified clay, unreinforced concrete, and gas-welded steel. Low-strength (less than 0.4 in. [10 mm] tolerance) pipelines include cast iron and vitrified clay with mortared or leaded joints.

*Valves and other appurtenances*. Flexible connections should be provided for valves and other appurtenance to reduce earthquake effects. A proactive preventive maintenance program for all valves and other appurtenances is also a critical element in any mitigation program. Such a program should include identifying critical valves, locating them in the field, and establishing location reference points. Valves should be inspected and repaired regularly.

*Repair*. Quick repair of broken mains and service lines is crucial during emergencies. Adequate stockpiles of such materials as repair clamps, sleeves, pipe, and valves should be maintained. This is especially important for materials that have long delivery lead times. All materials should be stored properly and protected from disaster hazards. Utilities should also maintain and have ready access to miscellaneous hand tools and debris-removal equipment.

## Control Systems

It is important to implement reliable monitoring and control systems. The monitoring system should gather information on pressures and flows. Where feasible, the control system should allow isolation of impacted areas.

## Interagency Connections

An interagency distribution system connection is a physical connection between water sources operated by adjacent water agencies (Boyle 1980). A typical interconnection consists of valves and a meter located in a simple in-ground vault or aboveground pump house. If the two water systems operate at significantly different pressures, a pump or regulator is needed at the interconnection to accommodate the pressure differential. A written agreement between agencies should specify the conditions of use, necessary notifications before use, construction and maintenance responsibilities, and operational procedures. Raw water transfer between agencies may also be considered.

## Administration, Transportation, and Communication

Administration facilities, equipment, and records are vital to the operation of a water system. Structures that house administrative functions should meet applicable building codes or retrofitted to meet minimum disaster needs. This will not only protect the structures, equipment, and records located inside but will help protect personnel as well.

Many of the specific recommendations for mitigating disaster effects found in the earlier discussion of treatment facilities apply to administration facilities. It is crucial to maintain backup copies of customer information, system maps and records, and the emergency plan at secondary locations. General facility and computer system (including control systems) security must be maintained.

Damage to the transportation system, including roadways and vehicles, can occur in many disasters. Mitigation actions include identifying alternative methods of access to system facilities, confirming that vehicle storage structures are disaster resistant and have auxiliary power, and maintaining adequate supplies of fuel and spare parts.

Communication disruptions due to phone and electric outages are difficult to mitigate. The most common alternative is to communicate using a two-way radio. Be sure to have enough radios and batteries. Mitigating effects on telemetry were discussed in the earlier section on treatment. Communication lines need to be open to other lifeline agencies and the mass media. Consider contacting amateur radio groups to provide emergency communications.

## REFERENCES AND SUPPLEMENTAL READINGS

American Society of Civil Engineers, Technical Council on Lifeline Earthquake Engineering. 1983. Advisory Notes on Lifeline Earthquake Engineering. New York: ASCE.

ANSI/AWWA D110. 2005. Standard for Wire-Wound Circular Prestressed-Concrete Water Tanks. AWWA.

ANSI/AWWA D100. 1997. Standard for Welded Steel Tanks for Water Storage. AWWA.

AWWA and American Society of Civil Engineers. 2012. (5th ed.) *Water Treatment Plant Design*. New York: McGraw-Hill.

AWWA and Japan Water Works Association. 1993. *Instrumentation and Computer Integration of Water Utility Operations*. Denver, CO: AWWA/AwwaRF.

Boyle, D.B. 1980. Interagency Connections: Insurance Against Interruptions in Supply. *Journal AWWA*, 72:4:192–195.

Denning, J. 1993. Seismic Retrofitting: Spending to Save. *Civil Engineering*, 63:2:48–51.

FEMA (Federal Emergency Management Agency). *Hazard Mitigation Assistance*. http://www.fema.gov/hazard-mitigation-assistance.

FEMA. *Hazard Mitigation Planning*. www.fema.gov/hazard-mitigation-planning.

USEPA (US Environmental Protection Agency). *Hazard mitigation assistance programs*. www.epa.gov/fedfunds/hazard-mitigation-assistance-programs-available-water-and-wastewater-utilities.

US Geological Survey. *Earthquake Hazard Program*. earthquake.usgs.gov/learn/topics/measure.php.

This page intentionally blank.

Appendix **A**

# Additional Resources for Emergency Preparedness and Response

Following are 10 "rip and run" style checklists to facilitate emergency preparedness, response, and recovery actions for water utilities:

- Incident Action Checklist - Drought (PDF) (8 pp, 2MB) | EPA 817-F-15-001, January 2015
- Incident Action Checklist - Earthquake (PDF) (8 pp, 997K) | EPA 817-F-15-002, January 2015
- Incident Action Checklist - Extreme Cold and Winter Storms (PDF) (8 pp, 987K) | EPA 817-F-15-003, January 2015
- Incident Action Checklist - Extreme Heat (PDF) (7 pp, 772K) | EPA 817-F-15-004, January 2015
- Incident Action Checklist - Flooding (PDF) (9 pp, 2MB) | EPA 817-F-15-005, January 2015
- Incident Action Checklist - Hurricane (PDF) (8 pp, 2MB) | EPA 817-F-15-006, January 2015
- Incident Action Checklist - Tornado (PDF) (8 pp, 1MB) | EPA 817-F-15-007, January 2015
- Incident Action Checklist - Tsunami (PDF) (8 pp, 1MB) | EPA 817-F-15-008, January 2015
- Incident Action Checklist - Volcanic Activity (PDF) (10 pp, 2MB) | EPA 817-F-15-009, January 2015
- Incident Action Checklist - Wildfire (PDF) (8 pp, 912K) | EPA 817-F-15-010, January 2015

This page intentionally blank.

# Appendix B

# Emergency Response Plan Crosswalk for Plan Review Foreword Section

1. Foreword/Preface/Introduction: Provides a foreword, preface, or introduction that explains why the plan was developed and how the plan is to be used. This section should explain that the plan is flexible enough to use in all emergencies and will facilitate response and short-term recovery activities (National Incident Management System [NIMS] emergency response plan [ERP] Element).
2. Plan Concurrence: Provides evidence that the assigned emergency agencies agree with how the plan describes their tasks. This may be in the form of a letter of concurrence or a sign-off sheet.
3. Letter of Approval: Provides evidence of a dated letter of promulgation or resolution from the governing board.

## PART I: BASIC PLAN

4. Table of Contents: Listing of where significant parts of the plan are located by page number and subsection of the plan.
5. Purpose: Describes the ERP purpose (NIMS ERP Element).
6. Scope: Defines the scope of preparedness and incident management activities necessary for the jurisdiction (NIMS ERP Element).

7. Authorities and References: Describes the ERP authorities and references (NIMS ERP Element), provides authorities for the plan and its development, and identifies the references used in developing the plan.
8. Situation and Assumptions: Describes the ERP situation and assumptions (NIMS ERP Element), summarizes the jurisdictional hazard analysis, and includes a description of potential hazards. This could be in a narrative with maps, schematic, or matrix indicating severity potential, affected population estimates, frequency, and geographical characteristics of the jurisdiction. This and other relevant information should be included to provide a rationale for prioritizing emergency preparedness actions for specific hazards.
9. Organization, Roles, and Responsibilities: Describes organizational structures, roles and responsibilities, policies, and protocols for providing emergency support (NIMS ERP Element).
10. NIMS-based Emergency Organization: Identifies agency roles and responsibilities during disaster situation, includes an emergency organization chart, and indicates how the jurisdiction fulfills the five NIMS sections (Management, Plans/Intelligence, Operations, Logistics, and Finance/Administration).
11. Emergency Operations Center Organization: Describes the roles and responsibilities of agencies and departments in the emergency operations center (EOC), including who is responsible for ensuring the readiness of the EOC.
12. Involvement of Special Districts, Private and Nonprofit Agencies: Identifies emergency responsibilities of special districts, private and volunteer agencies, and their roles in the EOC, Regional Emergency Operations Center, incident command post, or other emergency facilities.
13. Essential Facilities—Primary and Alternate EOC: Indicates the location of both the primary and alternate EOCs and what conditions would cause the alternate EOC to be activated.
14. Essential Facilities—Activation/Deactivation of EOC: Indicates how, when, and by whom the EOC will be activated and deactivated.
15. Essential Facilities—Alternate Government Facilities: Indicates an alternate seat of government to serve as government offices for performing day-to-day functions and a facility that could serve as an alternate EOC.
16. Essential Facilities—Americans with Disabilities Act: Identifies how shelter facilities, evacuation/movement, warning, and similar procedures accommodate the provisions of the Americans with Disabilities Act.
17. Continuity of Government: Provides persons by position to succeed key government officials and members of the emergency management organization. Also indicates the level and duration of authority these individuals would assume.
18. Vital Record Retention: Indicates how vital records are to be protected in the event of a disaster. Most data storage systems have a backup system. Identifies the system, archiving schedules, and who has responsibility for its maintenance.
19. Concept of Operations: Describes the ERP concept of operations (NIMS ERP Element) and includes the principles and methods used to carry out emergency operations, including the provision of emergency services by government agencies.
20. Notification and Mobilization: Describes how resources are mobilized and managed and includes methods to contact emergency response personnel during normal hours and after hours, which may be in the form of an alert list.
21. National Incident Management System Coordination Levels: Indicates how the jurisdiction coordinates between the different NIMS levels (field, local, operational areas, region, state), how information is exchanged, how and when multi/interagency coordination and unified command are used. The operational area agreement should also be referenced, and the plan should indicate who performs the operational area responsibilities.

22. Incident Command System (ICS): Indicates how the ICS will be used in the field. This should include the interface between the field incident command post and the EOC. It should also indicate methods of integrating state and federal field activities into local emergency management operations. It predesignates jurisdictional and/or functional area representatives to the Incident Command or Unified Command whenever possible (NIMS ERP Element).
23. Field/Emergency Operations Center Communications and Coordination: Indicates how the EOC will coordinate and communicate with field units, operational areas, regions, and other entities, including the use of the response information management system.
24. Field/Emergency Operations Center Direction and Control Interface: Describes the direction and control relationship between the field responders (ICS) and the EOC. This should include the reporting of pertinent information.
25. Field Coordination with Department Operations Centers (DOCs) and Emergency Operations Centers: Includes the use and coordination of DOCs and how they fit into the emergency management organization.
26. Mutual Aid: Includes a general description of the mutual aid system and processes.
27. Emergency Declarations: Indicates the purpose and process of emergency declarations (include samples).
28. Public Information: Includes pre- and post-incident public awareness, education, and communications plans and protocols (NIMS ERP Element).
29. Recovery Overview: Includes a general recovery concept of operations.
30. Recovery Organization: Provides a description of the recovery organization along with a diagram.
31. Recovery Damage Assessment: Describes the damage assessment organization and responsibilities.
32. Recovery Documentation: Describes the documentation process.
33. Recovery After-Action Reports: Includes the Office of Emergency Services after-action questionnaire.
34. Recovery Disaster Assistance: Describes the different programs, their purpose, restrictions, and application process. Includes public assistance, individual assistance, and hazard mitigation grant programs.
35. Administration and Logistics: Describes the administration and logistics of the ERP (NIMS ERP Element).
36. Emergency Plan Maintenance and Distribution: Describes ERP development and maintenance (NIMS ERP Element), indicates who maintains the emergency plan and what the process is, and details the schedules for modifications, revision list, distribution list, and who has responsibility for ensuring the plan is kept up-to-date.
37. Standard Operating Procedures Development: Ensures emergency response agencies develop and maintain standard operating procedures (SOPs) and indicates in the plan the relationship and purpose of SOPs to the plan.
38. Training and Exercises: Describes the training and exercise programs for the jurisdiction, including who has personal responsibility for the programs. Training should include ERP orientation, NIMS training, a full-scale exercise, and other trainings as deemed necessary.

## PART II: FUNCTIONAL ANNEXES

39. Functional Annexes: Contains functional annexes (NIMS ERP Element). Suggested annexes should address the five emergency management functions. They may be as simple as a checklist or as complex as to include function-based concepts of operation.

40. Management Section: Should include the following activities and responsibilities:
    - Overall EOC management
    - Public information officer assignment
    - Identification of a media center
    - Rumor control
    - Public inquires
    - Provision for public safety communications and policy
    - Identification of a safety officer
    - Facility security
    - Agency liaison
    - State/federal field activity coordination
41. Operations Section: Should include the following activities and responsibilities:
    - General warning
    - Special population warning
    - Authority to activate emergency alert system
    - Inmate evacuation
    - Traffic direction and control
    - Debris removal
    - Evacuation and care for pets and livestock
    - Access control
    - Hazardous materials management
    - Coroner operations
    - Emergency medical care
    - Transportation management
    - Crisis counseling for emergency responders
    - Urban search and rescue
    - Disease prevention and control
    - Utility restoration
    - Flood operations
    - Initial damage assessments
    - Safety assessments
    - Shelter and feeding operations
    - Emergency food and water distribution
42. Planning/Intelligence Section: Should include the following activities and responsibilities:
    - Situation status
    - Situation analysis
    - Information display
    - Documentation
    - Advance planning
    - Technical services

- Action planning
- Demobilization

43. Logistics Section: Should include the following activities and responsibilities:
    - Field incident support
    - Communications support
    - Transportation support
    - Personnel
    - Supply and procurement
    - Resource tracking
    - Sanitation services
    - Computer support
44. Finance/Administration Section: Should include the following activities and responsibilities:
    - Fiscal management
    - Time-keeping
    - Purchasing
    - Compensation and claims
    - Cost recovery
    - Travel request, forms, and claims

## PART III: APPENDICES

45. Appendices: Contains hazard-specific appendices (NIMS ERP Element).
46. Hazardous Materials: Incorporates or references the hazardous materials area plan requirements into the emergency plan.
47. Dams: If there are dams in the area, the plan should have, or reference, inundation maps that indicate what areas could flood, the time the flood wave would arrive at specific locations, and when the water would recede. Operational information necessary to carry out an evacuation of all potentially flooded areas should be indicated for each dam. This information, which is required for each dam, should include shelter locations, location of critical facilities such as government center hospitals, nursing homes, schools, and day care centers. Each dam evacuation plan should also indicate other facilities with large concentrations of disabled persons or persons who lack their own transportation or require special assistance.
48. Other Hazards Specific to the Jurisdiction: The threat of domestic terrorism has gained the interest of emergency managers in recent years. Other hazards could include damages from earthquakes or tsunamis in some coastal jurisdictions; some alpine areas may be prone to avalanches and some to volcanic activity. The ERP should address response activities that are specific to all hazards that pose a threat to the jurisdiction being evaluated.
49. Glossary of Terms: Contains a glossary of terms (NIMS ERP Element). Provides a glossary that includes all the terms used throughout the plan.
50. Resources: Identifies sources for materials and supplies internally and externally.
51. Contact List: Includes a list of agencies and personnel not internal to the organization but critical to emergency operations.
52. Supporting Documentation: Includes material necessary to self-certify compliance with NIMS. This should include evidence of training, planning, exercises, and performance.

This page intentionally blank.

Appendix C

# Federal Policies That Support Preparedness Planning

There are several presidential directives and public laws that affect emergency planning for water and wastewater utilities. Table C-1 provides an overview of national-level policies that support preparedness actions at the federal, state, and local levels and in collaboration with the private sector to effectively and efficiently prepare for, respond to, and recover from incidents, regardless of cause, size, or complexity.

**Table C-1 Policies and legislations that support preparedness planning**

| Title | Date | How It Relates to Utilities | Other |
|---|---|---|---|
| Safe Drinking Water Act | 1974; amended 1986, 1996 | Provides framework for utility to deliver safe drinking water; amendment<br>Summary of the Safe Drinking Water Act: http://www.fas.org/sgp/crs/misc/RL31243.pdf | |
| Presidential Decision Directive-63 | 1998 | Water sector designated as critical infrastructure; assigns USEPA responsibility to develop improvements to water infrastructure security http://www.fas.org/irp/offdocs/pdd/pdd-63.htm | |
| Stafford Act (as amended by PL 106-390) | 2000 | Authorizes the president to issue major disaster or emergency declarations in response to catastrophes that overwhelm state and local governments. Such declarations result in the distribution of a wide range of federal aid to individuals and families, certain nonprofit organizations, and public agencies. The forms of assistance authorized by the Stafford Act include temporary housing, grants for personal uninsured needs of families and individuals, repair of public infrastructure, emergency communications systems, and other forms of assistance. Congress appropriates money for activities authorized by the Stafford Act to the DRF, which is administered by the Federal Emergency Management Agency within the DHS. Appropriations to the DRF remain available until expended. Utilities can apply for disaster mitigation funding. | |
| US Patriot Act of 2001 (P.L. 107-56) | 2001 | Designates water as one of several "critical infrastructure" sectors, which are defined as "systems and assets, whether physical or virtual, so vital to the United States that the incapacity or destruction of such systems and assets would have a debilitating impact on security, national economic security, national public health or safety or any combination of those matters" (Sec. 1016(e)). | |

*(continued)*

**Table C-1 Policies and legislations that support preparedness planning (continued)**

| Title | Date | How It Relates to Utilities | Other |
|---|---|---|---|
| Safe Drinking Water Act amended by the Bioterrorism Act PL107-188 | 2002 | Mandates a vulnerability assessment for all public water systems serving 3,300 or more and requires submission to the USEPA. Mandates an emergency response plan informed by the findings of the vulnerability assessment and requires a letter of verification to the USEPA. Makes tampering with a drinking water system a federal offense with maximum civil penalty of $1,000,000 and/or up to 20 years in prison. | Section 1433 exempts contents of vulnerability assessment from public disclosure under the Freedom of Information Act. Provides for penalties for inappropriate disclosure of information by government officials. Sections 1434 and 1435 direct USEPA to review methods for contamination protection and response. |
| HSPD 3—Homeland Security Advisory System | | HSPD 3 created a Homeland Security Advisory System to inform all levels of government and local authority, as well as the public, about the current risk of terrorist acts. The system involves a five-level, color-coded threat condition indicator to correspond to the current situation. Agency-specific protective measures associated with each threat condition allow a flexible, graduated, and appropriate response to a change in the nation's level of risk. | April 20, 2011: This was replaced with the terrorism alert system. |
| HSPD 5—Management of Domestic Incidents | | Serves to enhance the ability of the United States to manage domestic incidents by establishing a single, comprehensive NIMS. The NIMS is designed to cover the prevention, preparation, response, and recovery from terrorist attacks, major disasters, and other emergencies. The implementation of such a system would allow all levels of government throughout the nation to work efficiently and effectively together. The directive gives further detail on which government officials oversee and have authority for various parts of the NIMS. HSPD 5 also establishes that utility staff are first responders and should be trained as such. | |
| HSPD 8—National Preparedness; superseded by PPD-8 | | Established policies to strengthen the United States' preparedness to prevent and respond to threatened or actual domestic terrorist attacks, major disasters, and other emergencies. The directive requires a national domestic all-hazards preparedness goal, with established mechanisms for improved delivery of federal preparedness assistance to state and local governments. It also outlines actions to strengthen preparedness capabilities of federal, state, and local entities. This is a companion directive to HSPD 5. Most utilities are one of the local entities that can receive federal assistance. | |

*(continued)*

**Table C-1 Policies and legislations that support preparedness planning (continued)**

| Title | Date | How It Relates to Utilities | Other |
|---|---|---|---|
| HSPD-9—Defense of United States Agriculture and Food | 2004 | Set mandates for the USEPA regarding water quality through surveillance and monitoring programs. The USEPA responded by launching the Water Security Initiative program, which focused on the development and deployment of water contamination warning systems. | |
| HSPD 21—Public Health and Medical Preparedness | | It is the policy of the United States to plan and enable provision for the public health and medical needs of the American people in the case of a catastrophic health event through continual and timely flow of information during such an event and with rapid public health and medical response that marshals all available national capabilities and capacities in a rapid and coordinated manner. Utility staff, as first responders, are among those with priority for flu vaccines. | |
| Support Anti-Terrorism by Fostering Effective Technologies Act of 2002 (Safety Act; part of the Homeland Security Act, Pl 107-296) | | Grants the secretary of DHS with broad discretion in determining whether to designate a particular technology as qualified anti-terrorism technology.<br>Bestows liability protections to providers and users of certain antiterrorism technologies and incentives for the development and deployment of these technologies by creating a system of risk and litigation management. | |
| Chemical Facility Anti-Terrorism Standards (Sec. 550 of DHS Appropriations Act of 2007) | 2007 | Directed DHS to issue rules establishing risk-based performance standards for the security of chemical facilities. Public water systems and wastewater treatment facilities as defined by Section 1401 of the Safe Drinking Water Act and Treatment Works as defined in Section 212 of the Federal Water Pollution Control Act, respectively, are excluded from this regulation. | |
| PPD-8—National Preparedness | | Aimed at strengthening the security and resilience of the United States through systematic preparation for the threats that pose the greatest risk to the security of the nation, including acts of terrorism, cyberattacks, pandemics, and catastrophic natural disasters. | |

*(continued)*

**Table C-1 Policies and legislations that support preparedness planning (continued)**

| Title | Date | How It Relates to Utilities | Other |
|---|---|---|---|
| Executive Order 13636—Improving Cyber Security for Critical Infrastructure | 2013 | Directs the Executive Branch to do the following:<br>• Develop a technology-neutral voluntary cybersecurity framework;<br>• Promote and incentivize the adoption of cybersecurity practices;<br>• Increase the volume, timeliness, and quality of cyber threat information sharing;<br>• Incorporate strong privacy and civil liberties protections into every initiative to secure our critical infrastructure; and<br>• Explore the use of existing regulation to promote cybersecurity. | |
| PPD-21- Critical Infrastructure Security and Resilience | 2013 | Advances a national unity of effort to strengthen and maintain secure, functioning, and resilient critical infrastructure. This endeavor is a shared responsibility among the federal, state, local, tribal, and territorial entities and public and private owners and operators of critical infrastructure (herein referred to as "critical infrastructure owners and operators"). This directive also refines and clarifies the critical infrastructure-related functions, roles, and responsibilities across the federal government, as well as enhances overall coordination and collaboration. The federal government also has a responsibility to strengthen the security and resilience of its own critical infrastructure, for the continuity of national essential functions, and to organize itself to partner effectively with and add value to the security and resilience efforts of critical infrastructure owners and operators.<br><br>PPD-21 replaces DHS Presidential Directive-7 and directs the Executive Branch to do the following:<br>• Develop a situational awareness capability that addresses both physical and cyber aspects of how infrastructure is functioning in near-real time,<br>• Understand the cascading consequences of infrastructure failures,<br>• Evaluate and mature the public–private partnership,<br>• Update the National Infrastructure Protection Plan, and<br>• Develop a comprehensive research and development plan. | |

DRF: Disaster Relief Fund, DHS: Department of Homeland Security, HSPD: Homeland Security Presidential Directive, NIMS: National Incident Management System, PPD: presidential policy directive, USEPA: US Environmental Protection Agency

This page intentionally blank.

# Glossary

The following concepts and terms in this glossary are used throughout the manual to discuss emergency planning.

**All-hazards** An approach for prevention, protection, preparedness, response, and recovery that addresses a full range of present threats and hazards, including malevolent attacks, natural, technological, and human-caused hazards, and other emergencies.

**Asset** An item of value or importance. In the context of critical water, wastewater, and reuse infrastructure, an asset is something of importance or value that if targeted, exploited, destroyed, or incapacitated could result in injury, death, economic damage to the owner of the asset or to the community it serves.

**Consequences** The immediate, short- and long-term effects of a detrimental natural, technological, deliberate or accidental human-caused incident or event. These effects include losses suffered by the owner of the asset and by the community served by that asset.

**Crisis management** For the private sector, crisis management is that transition from normal business decision-making process to a highly streamlined process aimed at containing the initiating event, maintaining essential operations, and recovery of normal business conditions as quickly as possible.

**Dependency hazards** Risks from supply chain disruptions. May include source to tap interruptions.

**Design basis threat** A plan that spells out the specific details of an attack that a utility should prepare to withstand.

**Emergency** An unforeseen or unplanned event that may degrade the quality or quantity of potable water supplies available to serve customers.

**Emergency response plan** Utility document that, at a minimum, defines procedures and roles/responsibilities during an emergency.

**Event** Impact to the utility's system that hinder normal operations on a wide scale (there can be several incidents in one event).

**Hazard** Something that is dangerous or harmful to life or property from a naturally occurring or human-induced disaster event. For example, the high winds of a hurricane or the ground shaking of an earthquake is a hazard. (See *All-hazards*)

**Human-caused disaster** A disaster caused by intentional or accidental acts. Human-caused disasters may be the result of employee error, transportation accidents, employee work stoppage, vandalism, civil disorders, terrorism, biological contamination, hazardous chemical spills, equipment failures, explosions and incendiary fires, or warfare.

**Incident** An unexpected interruption of normal operations.

**Lifeline concept** The infrastructure of networks and facilities that provides essential life and health services to citizens, businesses, and industry. Lifelines include water/wastewater, power utilities, transportation networks, communication systems, medical care facilities and equipment, fire and police facilities, and emergency operations centers.

**Malevolent act** Deliberate acts carried out by person(s) with the intention of causing harm to others.

**Natural disaster** A disaster caused by natural forces or events that create water utility emergencies. Examples include earthquakes, forest or brush fires, hurricanes, tornadoes or high winds, floods, and other severe weather conditions, such as freezing or drought.

**Preparedness** A continuous cycle of planning, organizing, training, equipping, exercising, evaluating, and taking corrective action in an effort to ensure effective coordination during incident response.

**Preparedness plans** Documents with instruction regarding preparedness and emergency response. These documents may also be referred to as emergency response plans, emergency operation plans, or hazard-specific plans. For the purpose of this document, the plan that ties together all preparedness plans for a utility is referred to as an emergency response plan.

**Proximity hazards** Collateral damages that a water system or facility can suffer from at outside sites in the proximity of the utility's assets.

**Risk** Exposure to the chance of harm or loss. Risk is a function of the likelihood of an unwanted event occurring, vulnerability, and the consequences resulting from the outcome of the threat.

**Risk assessment** A process to identify hazards and their relative probabilities of occurrence; to identify assets at risk; and to quantify the potential impacts of the hazards on assets at risk.

**Threat** Any indication, circumstance, or event with the potential to cause the loss of, or damage to, an asset or population. (See *All-hazards*)

**Utilities** Any public or private water and wastewater agency, and any water management entity.

**Vulnerability** An inherent state of a system that can be exploited by an adversary or impacted by a natural, technological, or human-caused threat to cause harm or damage.

# Acronyms

| | |
|---|---|
| AAR | After Action Report |
| AAR / IP | After Action Report / Improvement Plan |
| ASCE | American Society of Civil Engineers |
| ASDWA | Association of Drinking Water Administrators |
| BCP | Business Continuity Plan |
| CBWR | Community-based Water Resiliency Electronic Tool |
| CCP | Crisis Communication Plan |
| CDC | Center for Disease Control |
| CFR | Code of Federal Regulations |
| COOP | Continuity of Operations Plan |
| DHS | Department of Homeland Security |
| DRF | Disaster Relief Fund |
| EOC | Emergency Operations Center |
| ERP | Emergency Response Plan |
| FEMA | Federal Emergency Management Agency |
| HSEEP | Homeland Security Exercise Evaluation Program |
| ICS | Incident Comment System |
| MAA | Mutual Aid/Assistance Agreement |
| NIMS | National Incident Management System |
| NIST | National Institute of Standards and Technology |
| NGO | Nongovernmental organizations |
| PIO | Public Information Officer (sometimes referred to as Public Information Office) |
| SCADA | Supervisory Control and Data Acquisition |
| SMS | short message service |
| USACE | United States Army Corps of Engineers |
| USEPA | United States Environmental Protection Agency |
| WARN | Water/Wastewater Agency Response Network |

This page intentionally blank.

# Index

Note: *f.* indicates figure; *t.* indicates table

## P



## R



## S



## T

This page intentionally blank.

# AWWA Manuals

M1, *Principles of Water Rates, Fees, and Charges*, #30001
M2, *Instrumentation and Control*, #30002
M3, *Safety Management for Water Utilities*, #30003
M4, *Water Fluoridation Principles and Practices*, #30004
M5, *Water Utility Management*, #30005
M6, *Water Meters—Selection, Installation, Testing, and Maintenance*, #30006
M7, *Problem Organisms in Water: Identification and Treatment*, #30007
M9, *Concrete Pressure Pipe*, #30009
M11, *Steel Pipe—A Guide for Design and Installation*, #30011
M12, *Simplified Procedures for Water Examination*, #30012
M14, *Backflow Prevention and Cross-Connection Control: Recommended Practices*, #30014
M17, *Fire Hydrants: Installation, Field Testing, and Maintenance*, #30017
M19, *Emergency Planning for Water and Wastewater Utilities*, #30019
M20, *Water Chlorination/Chloramination Practices and Principles*, #30020
M21, *Groundwater*, #30021
M22, *Sizing Water Service Lines and Meters*, #30022
M23, *PVC Pipe—Design and Installation*, #30023
M24, *Planning for the Distribution of Reclaimed Water*, #30024
M25, *Flexible-Membrane Covers and Linings for Potable-Water Reservoirs*, #30025
M27, *External Corrosion Control for Infrastructure Sustainability*, #30027
M28, *Rehabilitation of Water Mains*, #30028
M29, *Water Utility Capital Financing*, #30029
M30, *Precoat Filtration*, #30030
M31, *Distribution System Requirements for Fire Protection*, #30031
M32, *Computer Modeling of Water Distribution Systems*, #30032
M33, *Flowmeters in Water Supply*, #30033
M36, *Water Audits and Loss Control Programs*, #30036
M37, *Operational Control of Coagulation and Filtration Processes*, #30037
M38, *Electrodialysis and Electrodialysis Reversal*, #30038
M41, *Ductile-Iron Pipe and Fittings*, #30041
M42, *Steel Water-Storage Tanks*, #30042
M44, *Distribution Valves: Selection, Installation, Field Testing, and Maintenance*, #30044
M45, *Fiberglass Pipe Design*, #30045
M46, *Reverse Osmosis and Nanofiltration*, #30046
M47, *Capital Project Delivery*, #30047
M48, *Waterborne Pathogens*, #30048
M49, *Quarter-Turn Valves: Head Loss, Torque, and Cavitation Analysis*, #30049
M50, *Water Resources Planning*, #30050
M51, *Air Valves: Air-Release, Air/Vacuum and Combination*, #30051
M52, *Water Conservation Programs—A Planning Manual*, #30052
M53, *Microfiltration and Ultrafiltration Membranes for Drinking Water*, #30053
M54, *Developing Rates for Small Systems*, #30054
M55, *PE Pipe—Design and Installation*, #30055
M56, *Nitrification Prevention and Control in Drinking Water*, #30056
M57, *Algae: Source to Treatment*, #30057
M58, *Internal Corrosion Control in Water Distribution Systems*, #30058
M60, *Drought Preparedness and Response*, #30060
M61, *Desalination of Seawater*, #30061
M62, *Membrane Applications for Water Reuse*, #30062
M63, *Aquifer Storage and Recovery*, #30063
M65, *On-Site Generation of Hypochlorite*, #30065
M66, *Cylinder and Vane Actuators and Controls—Design and Installation*, #30066
M68, *Water Quality in Distribution Systems*, #30068

City of Thornton
Water Treatment

www.ingramcontent.com/pod-product-compliance
Lightning Source LLC
LaVergne TN
LVHW082350200926
848481LV00012B/3524